ONE MAN
AND HIS
HOG

CW01498192

ONE MAN AND HIS HOG

THE STORY OF A PIG CALLED ALICE

PAUL HEINEY

First published 2019
This paperback edition first published 2025

The History Press
97 St George's Place, Cheltenham,
Gloucestershire, GL50 3QB
www.thehistorypress.co.uk

British Library Cataloguing in Publication Data.
A catalogue record for this book is available from the British Library.

ISBN 978 1 83705 062 8

Typesetting and origination by The History Press
Printed and bound in Great Britain by TJ Books, Padstow, Cornwall

MIX
Paper | Supporting
responsible forestry
FSC   FSC® C013056
www.fsc.org

The History Press proudly supports

Trees for LYfe

www.treesforlife.org.uk

EU Authorised Representative: Easy Access System Europe
Mustamäe tee 50, 10621 Tallinn, Estonia
gpst.request@easproject.com

Contents

1 Who Was Alice?

To call Alice 'just another pig' would be the gravest insult. She was far removed from the ordinary, the common-or-garden, the routine. She had qualities that elevated her above the commonplace members of that species. All pigs are special, as those who have kept them will tell you, but there was something about Alice that went way beyond. She had a profound effect on me. If anyone were to ask, 'who has influenced you most in your life?' my first thought would be Alice.

She was my friend for the best part of a decade. We talked to each other on a regular basis and she listened patiently. I tended and cared for her and in return she gave me comfort too. There is a deep kind of pleasure to be had in the company of a pig (especially Alice) and even more if, in a shared moment, you lean over the wall of her sty and stroke her back with a stick, because pigs like that. Alice would have rolled her eyes, if she could. Blissful hours can be spent just watching a pig, and before long you will find yourself in the same kind of meditative state of mind that mystics seek. There is much medicine prescribed to treat a weary mind, but if doctors would recommend an hour a day talking to a pig, what a boost that would be to our spiritual health.

Much praise has been heaped on the pig for the multiplicity of its gifts. We know from countless writers that a human in need of food can use every part of the pig 'bar the squeak'. From their skin to their hearts, there is always something to be found that, with a little cooking skill, will make a fine feast. This much is well known.

But during those quiet moments that Alice and I shared, I often wondered if this pig had more to offer than just her flesh. Having watched how pigs behave – and in particular how Alice led her life and reared her offspring – I came to believe that, here before us, in every field and every sty where pigs are kept, there are lessons that we might learn to help *us* lead an improved life. Is it possible that pigs have a philosophy? Could it be that they have got life worked out to such an extent that their contentment is complete, and their troubles are few? I believe it to be so, and in the years during which I knew Alice I tried to unravel what this secret might be.

She was the only pig with whom I could hold a genuine conversation. It took time, of course, but soon I learned how to ask if she was happy with her sty, comfortable in her straw or generally at peace with the world. And if it was me who was going through a stressful patch – a common state of mental affairs for farmers large and small – then it was she who could calm me. In fact, she had almost medicinal potential as a soother of the fevered brow.

Alice inspired me in many ways. She forced me to think more deeply than ever before about the working relationship between a farmer and his animals, and what it tells us about ourselves. She also served to bring into sharp focus the damage that has been done to our respect for farm animals in the relentless pursuit of food that must be profitable, whatever the cost to an animal's dignity. Some people will look at a pig and see only chops, where I observed deep truths.

When we talk about Alice, we are speaking of a figure who was a true giant amongst her generation. She captured the hearts and imaginations of thousands who had never even met her through my writings in *The Times* newspaper over twenty years ago, where she was often mentioned, and my recollections here are drawn from my diaries at the time.

As pigs go, Alice the Large Black pig was as influential a sow as ever lived and when she died, she was mourned the length of the country. She was the people's pig. If ever a pig had greatness thrust upon it, it was Alice. Humbly born and expecting no more from life than the drudgery of rearing of piglets, Alice accidentally found fame. Call it charisma, call it star quality, Alice had it from the tip of her slimy snout to the very end of her curly tail. Once, from the far side of a crowded Oxford Street in London came the cry, 'How's Alice?' She was known everywhere. A rock star would have been jealous of Alice's mailbag, yet she took it like the lady she was and was patient and gracious with all enquiries. What a shame she didn't live to see the growth of YouTube, for she would certainly have been an influencer to be reckoned with, and certainly more intelligent than some. Now and again she showed a certain impatience with time-wasters: no disgruntled duchess ever gave a look more thunderous than that given by Alice, the Large Black pig. But, on the whole, I was lucky. With all that public attention she could have turned into a monster, but she remained to the very end the sweetest pig in the world.

By the way, don't imagine this is going to be one long, drippy lovelorn tale. If you think all this love and affection that I have been hinting at so far is going to be maintained, think again. She could be stubborn, abusive, violent and a complete bitch, all within the space of five minutes. No petulant rock star could outdo Alice when it came to temper. This relationship, believe me, was up and down like a yo-yo. But that, of course, is all part of the mystery of a pig called Alice.

A Quick
2 History of
Pigs

Yʏou had only to look at Alice to realise that who-
ever invented the pig did a damned good job.
There are estimated to be 800 million of them
now, honking and squealing in every corner of the
world, although the loudest din will come from China,
which has over half the world's pig population. That's
a lot of piggies. And they produce a mighty mountain
of meat – 60 million tons of pork a year in China alone,
making it the world's most widely consumed meat.

It is a remarkable success story for a creature with a
miserable beginning. The wild boar, from which our
much friendlier Alice is descended, was not the sort of
creature to meet on a dark night, nor one to find yourself within snapping distance of. You couldn't possibly
be fond of a wild boar. It's a surly, bristled creature with
mean eyes and a threatening snout, and the sharpest of
teeth that could rip you apart with ease. Nevertheless,
the police in Logroño, Spain, have one as their mascot
and take it around on a lead, like a dog. It is far from
clear whether it is the armed officer or the long-haired
boar with a body the size of a muscular pony that is
more likely to deter crime. It would be those evil tusks
that would do you most damage, assuming you hung
round for long enough, which you wouldn't because
the boar has an ability to threateningly raise its fur along
the length of its spine till it looks like a dragon from a
children's book. One glimpse of that and your courage would surely evaporate. In fact, where wild boar

16

hunts are written of in Scandinavian texts and Anglo-Saxon writings, they are described as being for only the truly courageous. Remember also that in the list of the Labours of Hercules, catching the Erymanthian Boar comes fourth. Assuming his tasks were in ascending order of difficulty, he would have already tackled a monster lion, a serpent and a hind with golden antlers before getting round to this wildest of pigs. Incidentally, his next task was to clean the Augean stables, which might be thought of as a bit of a doss after all that wild animal grappling. In fact, the stables, which housed a thousand cattle, had never been cleaned in thirty years. The inventive Hercules re-routed the flow of two rivers to eventually wash them out and thus avoid himself a truly Herculean effort with a muck fork.

But how do we get from the feared wild boar to the tractable, kindly creature we know and love – the farmyard pig? What happened to that vast family, of which Alice was a part, that reformed the wayward boar into the domesticated pig? When did those nasty pieces of sharp-toothed work transform themselves into cuddly Miss Piggy? How did fear and loathing of a species end up with the very same being beloved by generations of children, most recently through Pigling Bland, *The Sheep-Pig*, and, of course, Wilbur from *Charlotte's Web*?

It all started badly. For example, take what was known as the 'Irish Greyhound Pig', which was very similar to the earliest of pigs. They had long legs and great strength at their back end and were said to be able to jump over a pony if not a five-barred gate. They were, however, bony and not covered with much meat, carried coarse hair and had pendulous wattles hanging from their throats. It was a direct descendant of the pigs that had roamed the forests of Ireland since pre-historic times. Although now extinct, an Englishman, Sir Francis Head, came across similar animals in Germany and wrote:

> As I followed them this morning, they really appeared to have no hams at all; their bodies were as flat as if they had been squeezed in a vice; and when they turned sideways their long sharp noses and tucked-up bellies gave to their profile the appearance of starved Greyhounds.

The *Rural Cyclopaedia*, written by Revd John M. Wilson in 1848, pulled no punches and the author showed little Christian compassion to this clearly indifferent animal:

> It is ugly, bony, razor-backed, lank, coarse, greedy, a voracious eater, a most unkindly feeder, and exceedingly difficult to fatten; and even in its best state, it yields pork and bacon far inferior to those of almost all the improved breeds. Its best recommendation

is that it has disappeared from all comparatively enlightened parts of Ireland and is rapidly disappearing from even the remote and most neglected districts. Yet, with unutterable absurdity, it is still preferred to every other hog by a few of those half-mad antiquity-loving peasants who can see nothing but horror and ruin in any kind of deviation from the practices of their fathers.

The long, slow transformation of the pig started around the time of the early Bronze Age (2000 BC) when domestication kicked off with the intention of dissuading the ever-greedy pig from destroying every crop it could find. Having lived in an unconfined way since their creation and able to wander and do damage where and when they could, teaching them to live confined lives was an optimistic task for an ancient stock-keeper.

But it worked. Fifteen hundred years later, pigs were kept and gorged on by the Romans, who made pork their principal form of meat. But the pig as a successful commercial animal, rather than just a casual source of food, is a much more recent development. Robert Bakewell, born 1725 and described by the *Breeder's Gazette* as 'the patriarch of animal breeders', had made a major contribution to the development of cattle and sheep by breeding in a methodical way, although his influence on the pig was insignificant except upon a breed known as the Small White.

Bakewell preferred to work in secret, so exactly what he did with the Small White we shall never know, except to say that as a breed it is now extinct. Some of his breeding resulted in pigs that were described at the time as 'rickety' or 'fools'. However, his work with Leicester sheep and Longhorn cattle led to a wider understanding of how desirable traits in animals could be preserved and enhanced through selective breeding, which was a major breakthrough in animal husbandry. But perhaps his success with pigs suggests that they might have been cleverer than him. I have always detected a certain arrogance in pigs and it is quite possible that Bakewell's piggy subjects simply thought that they could not be improved upon and so ignored his experiments. By behaving like 'fools', they may have been making a fool of him.

The problem with the Small White, and all other pig breeds around that time, was that their keepers were not farmers but aristocratic hobbyists whose main interest was taking them to shows, winning cups, and being proud of them. They wanted to be seen owning pigs that were eye-catching: fatness was all that mattered. A glance at the engravings of the time shows pigs of the most hideous conformation, and a modern health adviser would surely be gasping as they struggled to speak the words 'morbidly obese'. And they were – fat as pigs. The result was that they hardly appealed to any palate, no matter how much the poor consumer might have enjoyed a bit of fat on their meat. In fact, it was more a case of a bit of meat on their fat.

To make it even more difficult for them to hold on to a slender frame, pigs became a convenient method of waste recycling. They were the first consumers of junk food. The big London breweries at the end of the eighteenth century fattened no less than nine thousand pigs a year on brewers' grains and all the other waste

that came from the rapidly expanding brewing business. They were useful too to the dairy farmer who had to dispose of whey, the by-product of cheese making. Boy, did pigs grow fat on that as well.

Fatness trumped everything when it came to pigs. The lashings of lard that surrounded pig meat were colossal by any measure, and the old wild boar suddenly found that it had been transformed over a couple of thousand years into a ridiculed creature, and not a productive or nutritious part of the army of animals that farmers reared to make themselves a living and provide us with food. It is probably from this time that the use of the word pig as an insult took hold. Anyone who was lazy, filthy or greedy was now called a 'pig'. Gluttons were said to 'eat like pigs' or 'have the table manners of a pig'. These days, in urban jungles, wayward youths will call the police 'the pigs', meaning they are seen as the lowest form of life. Middle-aged pink-faced men of conservative views are now being called 'gammons'; the creators of this insult presumably being unaware that the gammon is one of the pig's sweetest and richest gifts and to be revered. But it all goes back to those selfish, hobbyist aristocrats, probably fat themselves, who believed that, when it came to pig, big was beautiful and that was all that mattered. So horrified was the Revd Henry Cole that he wrote to the Royal Agricultural Society in 1852 having observed the pigs at a show in Brighton:

My pain arose from witnessing animals unable to stand, and scarcely able to move or breathe, from the state of overwhelming and torturing obesity, to which they had been unnaturally forced. Nor could one of the unhappy prized pigs, or of those exhibited to contest the prize, stand for one moment. And, if punished by being partially raised to gratify the brutish curiosity of a few, they laid themselves down, or rather fell down immediately. And it appears, that several died in their cart-conveyance to the scene of this cruel and unchristian exhibition.

In the nineteenth century, breeders began to realise that by judicious mating of breeding stock drawn from a wide range of differing breeds of pig, a more acceptable and leaner kind of meat could be produced. All of a sudden, the pig found itself in the ascendency. Gone were the old beliefs, best expressed by William Cobbett in his *Cottage Economy* when he wrote of the pig:

If he can walk two or three hundred yards at a time he is not well fatted. Lean bacon is the most wasteful thing that a family can use. In short, it is uneatable except by drunkards who want something to stimulate their sickly appetites.

I wonder what view Cobbett would take of a modern supermarket display of bacon, bred for minimal fat, a sliver of it twixt meat and skin, so slender that a magnifying glass might be needed to spot it? Cobbett was right, though, but only up to a point; bacon or pork with no fat is a tedious and flavourless kind of meat to eat. But there are limits, and perhaps a pig that is so fat that it can't get from one side of field to another without a lie down to get its breath back, might not produce the healthiest of foods.

And what has any of this to do with a pig like Alice, except that, clearly, she was a pig herself? She was of a breed called the Large Black, doubtless a creation of the genetic meddlers like Bakewell and those who came after him. And yes, she had a certain stoutness about her, I can't deny it, although I would never have dared to mention it in front of her – pigs can take offence too. She is truly part of that long history going back to the wild pigs of ancient times and a milestone on the long march of the pig's progress. Although much has been written about the history of the pig, the husbandry of them, the commercial exploitation and the cooking of them, little has been recorded about what it is like to live with a pig, and certainly not with a pig of a forceful personality like Alice. Alice was a treasure; an independent, strong-minded soul, but generous of spirit. In this book are the lessons she so selflessly taught me.

3 My First-Ever Pig

To understand how Alice the pig first came into my life, you must appreciate a very strange direction that my life had taken. Somewhere down the line, for reasons I still do not fully understand, I had become completely besotted with a totally romantic idea that I would become a farmer. Me? A farmer? Crazy idea. I was hardly a gardener, let alone anyone capable of expanding a few basic skills in the vegetable patch onto the scale needed to grow a decent crop of anything. I couldn't tell wheat from barley, a tup from a gilt or one end of a plough from another. But once the notion had been planted it seemed to flourish, watered and fed by ambition way beyond my skills. I simply had to do it. Something, someone, was dragging me down one of life's new pathways and I knew I must follow.

Just to add another layer of complexity to an already fanciful project, I decided that it would be a small-scale model of a working Victorian farm. I was going to deliberately turn back the clock, drag history to the present day. It would have no tractors but instead employ carthorses; the Suffolk Punch would be my choice, being the native breed of the county in which I was living.

To try and gather some skills and get my writer's soft hands used to life with a few more scars and callouses, I befriended a couple whose skill with working farm horses was legendary. For one testing year I laboured alongside them, learning how bleak and tough a horse-drawn farming life could be. But a year's immersion in the drudgery and disappointment of farming life did nothing to dissuade me from the course I had set myself. In fact, the worse it got the more I relished it. When the day's task was to take a pitchfork and clear out an entire winter's hoard of muck and rotting straw from a cattle yard, I attacked it with an energy which I never knew I had. A day spent ankle deep in animal filth became a gift.

For the rich experience my new-found friends and mentors, Roger and Cheryl, had given me on their farm, for the wisdom they imparted and, most importantly, the philosophy that saw them through the emotional see-saw that is farming life, I decided to buy them a present.

It was going to be a pig.

Apart from the Suffolk horses, they kept all the rare breeds once common in East Anglia – Red Poll cattle, Suffolk sheep (now a major commercial breed). These were known as 'the Suffolk Trinity'. But to this trio should really be added the Large Black pig.

This breed was once the pig of choice for East Anglian farmers. They were prized for being docile, and for being able to convert food to meat with enormous efficiency. It was true that they carried more fat than was acceptable in the late twentieth century, but that didn't make them any less of a noble animal. And who's to say that 'fat' won't have its day once again?

Perhaps the Black Pig is just biding its time. For reasons that are not clear, these pigs thrived at opposite ends of the country. They were found in large numbers in Essex, but equally in Devon and Cornwall. Yet hardly anywhere between. The Celtic influence of the West Country is hardly like the Anglo-Saxon culture that the Essex pigs were accustomed to. A pig travelling between the two must have had to adopt a new mindset somewhere near Reading.

If you have not seen a Large Black pig before, then you are immediately struck by its, er, largeness and blackness. While standing still in the landscape they might easily be confused with a hovering thunder-cloud, or when moving at speed – usually at feeding time – for a freshly fired cannonball. Underneath it all, of course, they are little different to a common pink pig and in my experience, when it comes time to put a piece of 'black' pork in the oven, it looks hardly any less pink than from any other breed of pig. Much of their colour comes from their hair, which is thick, black and wiry. I have had to persuade several pork eaters that meat from a Large Black pig does not look as though it has been painted with tar. Incidentally, the hair on their ears is extra-long and tough; it's highly sought after by fly fishermen, as it's said to be best for catching trout.

The other distinctive feature is their ears, which are called 'lop-ears' and hang heavy down the sides of their heads, not only protecting themselves but also blinkering their eyes. That is why they tend to throw their heads around, like long-haired girls do, in order to get a better view of the world. But from a farmer's point of view, I soon learned, a lop-eared pig is a great asset when it comes time to capture one of them. Their inability to see you approaching if you come at them from behind means that you can at least get within touching distances before that twitching snout senses your presence and the pig shoots off like a bullet from a gun.

Research was now needed if my idea of giving my friends a pig as a thank-you present was going to work. Not only did I have to find a pig, buy it and deliver it, I needed a little education in both the ways and looks of pigs, especially the Large Black. If I was going to buy my friends a good example, I needed to know what one would look like. The ways of pigs, I decided, would only be learned in the hard school of experience, not knowing at the time how testing that would prove to be. But with regards to looks, the textbooks of the early and mid-twentieth century, written when my adopted style of farming was still in vogue, were in total agreement about the Large Black pig – it was of limited use and not highly prized by butchers. This was a bit of a blow. It was branded a 'dual-purpose' pig, which meant it was suitable for the production of both pork and bacon. A bacon pig is leaner, and a good bacon pig is longer in the back to provide more rashers. A pork pig, on the other hand, carries more fat, which makes for tastier meat. It therefore becomes clear that anything described as being good at both jobs is going to be something of a compromise. On the plus side, in a typical book of the era, I read that, 'they mature quickly and are hardy and docile.' That's what I wanted to hear, and it was especially a relief to read 'hardy', since my pig husbandry skills were minimal and any creature I owned was destined to be self-managing up to a point, at least until I had learned a lesson or two.

But surely the Large Black can't be all that bad a pig? It has survived when many noble breeds have fallen at the trough. We no longer have the Cumberland Pig, extinct since 1960, although we still relish Cumberland sausages. Are they now fake? The Dorset Gold Tip, a pig with an orange tint, disappeared into a golden sunset in the 1960s; and the Lincolnshire Curly Coat, with its white, frizzy hair to ward off the chill winds of the Lincolnshire Fens, went about the same time. There was even a Small Black, known as a Suffolk Pig, which would have made it ideal in my imagined East Anglian farmyard. However, it was described as having a 'delicate constitution', which may have been its downfall.

In truth, the problems with all these breeds was their stoutness. They were mighty fat at a time when eating fashions were changing and leanness both on the plate and round our waists was becoming desirable. They survived the Second World War, just about, and served us proudly by putting energy-giving fat on the plate. But by the time the 1960s were swinging and farmers were gearing up for one of agriculture's more lucrative periods, tradition counted for less and the accountants' views prevailed. A few taps on those newly invented calculators and it became clear that these old breeds were less good at converting their feed into saleable meat, and so off they went, grunting into history. I was determined I would do my bit for the Large Black pig. Such a glorious creature is it that it *had* to survive.

I had no idea where you went to buy a pig, let alone a rare one. The internet was not quite upon us, otherwise I daresay it would have been quite easy. But back then, in the dark ages of over twenty years ago, you had to ask around. You spoke quietly, asked for the word to be put about, begged to be pointed in the direction of a contact who might facilitate an introduction.

'What is it you're after, then?' someone would ask in response to a softly spoken telephone call. The accent was usually very rural, the line often faint, as if speaking from a different age.

'Well, what I'm looking for …' I had rehearsed the next bit and had no real understanding of what it meant. 'I'm looking for a Large Black in-pig gilt.' I heard myself sounding silly; my slightly posh voice against his weathered tone. A gilt, I had taught myself, is a juvenile female pig who has yet to farrow, but in-pig meant that she was pregnant. So although I appeared to be giving my friends a single pig, I was in fact giving them possibly a dozen.

An intake of breath on the other end of the line. 'I might just know where there is one. I might just know.'

But the vital detail as to where this might be, this magical place where black pigs were born, was not yet revealed to me. Perhaps I had to prove myself first. It could not be rushed.

A couple of days later, a message came through. I was to head for the Norfolk borders. 'Mr Churchyard will do yer proud, he will,' said a voice, like a spy, passing secrets. 'Got a few good 'uns, he 'as.'

Supposing he did, and assuming he'd sell me one, exactly how do you move a pig? Can you put it on the back seat of the car, with a belt around it for its safety? I thought probably not, although it might have made a pleasant distraction for other motorists when they drew alongside me at a traffic light. No, I needed a trailer.

I had already started to amass a collection of vintage farm machinery, although to call it 'vintage' gives it a legitimacy it did not deserve. It was mostly a collection of the finest dilapidation; the sort of stuff that even scrap dealers would shun. But it worked once, why wouldn't it work again? Foolish boy, I was. Amongst this collection was a contraption that I decided would make a fine, if smallish, livestock trailer. It was ex-army, although it wasn't clear *which* world war it had fought in (probably the First). It had a drop-down ramp at the back, and it was covered in canvas to keep the rain out. Someone thought it had been used to carry ammunition, which might have made for a strong floor and suspension. The question was, might it carry a potentially explosive pig?

I used the word 'explosive' because Mr Churchyard had assured me the pig he had in mind for me was certainly 'in-pig', or pregnant. Mmmm. The journey on which I was to take her might last a couple of hours – what if the stress of travelling brought on her labour? How would the trailer cope? I checked it over. The tyres looked reasonable, no better than that. And there was a certain amount of rust bursting through the military green paint, but surely it would be strong enough. Optimistically, I hooked it up to the car, took a deep breath, and headed for Norfolk.

Thirteen Pigs
4 for the Price of One

A polite Mr Churchyard, a pig breeder of high repute, took me into a large barn where amongst a wealth of ordinary looking pink pigs, were a clutch of black ones. We leaned over the half door. I looked, then looked again. Was there some mistake? Apart from the sow, a clutch of wriggling little black creatures, like animated chunks of liquorice, were scampering in the straw. Some were already using their little snouts to root out any morsels that might be hidden beneath it. Others were nudging their mother's teats in the hope that a trickle of milk might flow.

I looked at Mr Churchyard and he saw the puzzled look on my face.

'Ah,' he said. 'Yes, she pigged a couple of nights ago. Had a 'running' service, see. I wasn't quite sure when.'

A running service happens in many human relationships, often on a drunken Saturday night. The male and female are enclosed in the same sty at a time when the farmer thinks the sow might be on heat, and the boar takes his chance when he can after, presumably, 'running' after her a bit – that's a running service. 'Self-service' might be a more appropriate way of describing it, more of a running buffet. As voyeurism is not a pig farmer's habit, it is generally not possible to know precisely when the coupling took place.

But the result was plain to see. I counted thirteen piglets, scampering around, new to world, looking for food and fun. Mr Churchyard reached for a cardboard box. 'You can put 'em in here. They'll go in the boot of your car.'

It soon became clear from watching that sow being parted from her piglets that a mother pig, freshly delivered, appears not to give a damn about any of her piglets. As we picked them up one by one and dropped them into a cardboard box, often chasing them before capture, she took not the slightest interest, unless in the scrabbling a morsel of missed food was revealed and then it had all her concentration. But hadn't I read that pigs were perfect mothers? Was that not the case?

Piglets captured, the sow now had to be shifted from sty to trailer. I reversed up to the door and then Mr Churchyard, a lifetime's experience of pig handling behind him, deftly used a couple of boards to limit her field of vision and allow her only a clear view ahead. Pigs, apparently, prefer to stick to the straight and narrow and only venture where they can clearly see. Blinkered by a board either side of her head, aided by a few scraps of food laid up the trailer ramp, she was soon inside. It looked so easy, but I wasn't fooled. This was pig handling at a very advanced level by a man who had a lifetime of keeping pigs, and I guessed it would be a long time before I would be able to make such a speedy transfer.

Swift and positive action seems to be the key. No sooner had that gloriously rotund backside disappeared into the inner darkness of the trailer, than we grabbed the ramp, raised it sharply, and put a chain across it. I don't think she even noticed. The box of piglets, looking like a selection of Black Magic chocolates, went into the boot of the car and I set off on the most nerve-wracking journey of my life. What happened if the trailer wouldn't take her weight? The tyres – which had looked quite healthy – had breathed something of a sigh when the sow's full weight came on them and looked perilously flat. The floor, which I had assumed was stout and strong, seemed to give under her weight. What if she fell through and ended up making her way down the A140 under her own steam? What if the piglets got together and formed an escape plan? I stopped at a cafe for the quickest cup of tea I have ever drunk, dared to peep under the tarpaulin that covered the trailer and saw that the sow was flat on her belly, seeming to have hardly a care in the world. Then I sheepishly opened the boot. No escapees; all fast asleep and snoring.

Mmm, I thought. This pig-keeping business isn't as tricky as I imagined.

A Pig –
5 the Perfect
Present?

My friends were delighted and that fat sow may well have been the most successful present I have ever given anyone. It was remarkable how she accustomed herself to her new surroundings. On arrival at her new home, when we dropped the gate of the trailer she took not the slightest interest. We called her and, although her ears twitched a bit, none of her bulk shifted. She was like an old lady who had settled herself in first class and wasn't going to budge just because someone had shouted 'All Change!'

After a while, we grabbed a yard brush and give her a decent prod on the vastness of her backside. This is possibly not the way to treat a newly delivered mother of thirteen but it did the trick. She scrabbled to her feet, the trailer rocking from side to side, and she raised first her front and then her back. The springs sank as she bent herself double to turn around so that she was now facing the exit. I thought she might make a run for it but she just stood there, grunting in a dissatisfied kind of way, trying to work out what this new place was. She raised her head, sniffed deeply, looked from side to side – but didn't move.

The bucket was sent for. A few morsels of food were scattered on the ramp leading, like a paper chase, onto the farmyard. Before she could actually see them, her nose clearly registered them and she bowed her head with a determination to reconcile the scent of food with its whereabouts. Then she got it. Like a dog after aniseed, that snout hit the ground with the force of a bullet and a combination of tongue and teeth had that pig feed down her gullet faster than a hungry teenager presented with a pizza.

Slowly, and with great attention, she followed the trail of food that led to her new home, which was a sty littered with straw and lit by a single lamp to give warmth to the piglets. Once she was inside, we closed the door behind her and went to the car to get the litter. They were still asleep. With great care we lifted each of them, only the size of the palms of our hands, and put them carefully beside her. She was not in the slightest bit interested. Top of her agenda was seeking out the next meal for herself, not motherhood. The piglets staggered around, wobbly on their feet, defeated by any lumps in the straw over which they couldn't clamber. Eventually, though, the sow gave up on her search for food and satisfied her only other desire: to go to sleep. She fell onto her belly with a thud, as if she had collapsed, just missing a couple of wandering piglets that had taken the foolish steps of standing beneath her. We held our breath until they emerged, unharmed, on the other side.

Within seconds she was asleep, eyes closed, gone to the world, entirely untroubled by her new home and surroundings.

But the piglets weren't, for in this prone position the sow's battery of teats was revealed. It was like opening time at a pub, when the cloths are taken off the pumps and the first pints are drawn. They latched on, one by one, squabbling a little, but eventually settling down to the serious business of gorging on mother's milk. I have never seen a more contented sight.

My friends looked on. 'Bloody good sow, ain't she?' said Roger. And we all agreed. She was indeed a bloody good pig. And then he said, 'When these piglets have a got a few more weeks on their backs, you'd best be 'avin' one.'

I wasn't expecting that. I didn't know what to say.

'If you're goin' to 'ave a farm, you've got to 'ave a black pig,' my friends insisted, laughing their heads off.

And so, some months later, I went back to that farm to collect a pig of my own.

And that was when all the trouble began.

At this stage, I had no farm at all. But I now had a pig. We were deeply enmeshed in the grinding tedium of the land-buying process, hoping and praying we might afford what was our dream patch of soil. It was perfect in every respect. Not too large, remembering that I intended to work it with horses and not tractors; it still had traditional stabling intact; and, most importantly, a row of three luxurious pigsties built in Victorian times, when even pigs were entitled to a bit of decent architecture around them. But, for now, my longed-for farm had to remain a hope, a dream, an image I carried with me through anxious nights, waiting for the bloody lawyers to do their tedious business. None of this was of any interest to the pig that was soon to become mine. All she needed was a new home, and I had to find one for her. It was tempting to walk her into the solicitor's office and ask if he might look after her while he finished off the soul-destroying legal details that were keeping us all on tenterhooks. I think it might have hastened the process.

In the end, a kindly neighbour gave my pig a temporary home. They kept a smallholding with goats, fruit trees and the best tended vegetable beds I had ever seen. They were the sort of sensible people who saw the housing of this pig as an honour, and in return for their kindness I decided that this sow would be called Alice, which rhymed with their surname. That's how she got her name. So, Alice it was from now onwards, and somehow it suited her.

One thinks, of course, of Alice in Wonderland; and there was indeed something girlish about Alice the pig, an innocence, but a curiosity as well. She was hardly six months old, but she already had a habit of flinging her head skywards so those large, floppy ears dropped away from her eyes and she could get a better view of the world, and from the way she peered and observed, clearly she saw the world as some kind of wonderland that had to be investigated. Yes, Alice was the perfect name for her.

It was nearly Christmas and, being on the east coast of England, we were feeling the deep chill of the east wind. In her temporary home, where our neighbour's goats had once been housed, straw was being spread deeply across the floor to make a bed fit enough for a princess. The date was set for her collection, a few days before Christmas Eve, and there was much excitement and anticipation as I reattached that crumbling old trailer to the car and went off to collect her.

Expecting something of a struggle, I was surprised by how eager she was to begin her new life. It was as if she too was sensing some kind of turning point in her life and was facing it with typical determination. I loaded her up, teasing her with morsels just as I had with her mother. She had certainly inherited a similar appetite. It was an hour's drive to her new home, and as I was to confirm many times over the following years, every minute with a pig in a trailer is an anxious time. I have never looked in my side mirrors with such frequency. My big fear? That the pig would have found her way through the tarpaulin cover, made a jump for it, and would be happily careering along the A12, snout skywards, tasting the air and gobbling up the scent of freedom. In fact, I think she slept the whole way, for when we eventually arrived and opened the trailer it was difficult to see her in the fading light. She had simply buried herself deep in the straw, hiding from the world, like someone on a lengthy journey on a National Express coach.

I should add that all this had been conducted in great secrecy. It was Christmas, remember?

And what might a thoughtful husband wish to give his wife for Christmas? Especially a wife who was proving totally supportive of her husband's mad scheme to become a farmer and had never said a doubting word, even though the family's finances were being stretched as tight as knicker elastic round a barrage balloon.

That was when I decided that I would give her Alice. It would be her Christmas present. The first bit of livestock on the farm would be hers. What an honour! Or, at least, that's how I hoped she would see it. Had any woman had such a Christmas present before? How could she not like it? Wouldn't anyone want to own a pig like Alice? It might be a bit of cliché, but doesn't something 'expensive, black and sexy' form the basis of a decent present for a woman, in some circles at least?

So many doubts lurked. What if my wife was repelled by the sight of this pig, wouldn't have anything to do with her? What might I do then? How do you rehome an orphan pig? I could already see that Alice was a thing of beauty, and early indications were that she had been blessed with more character than most, but often such things are in the eye of the beholder. What if my wife's view was less rosy than mine?

With the children swaddled against the cold, we walked down the lane on Christmas Day like the kings seeking the newborn. We had brought Alice

gifts in the shape of scoops of pig nuts. We nervously looked over the stable door. This was the closest the children had ever been to a pig. Alice sensed she was being watched. Do pigs have a sixth sense? No, they have far more than that – they have dozens of senses that we cannot understand. Pigs can get to their feet with great urgency, if they can be bothered, and Alice was swift to present herself in her full glory, eventually rising to her feet with a weary groan. The children laughed when I told them she was called Alice, and our six-year-old daughter said there was a girl called Alice in her class at school and she was fat too. Our son remained thoughtful. But my wife was overjoyed and the relief spread over me like melting pork dripping on hot toast. Alice was now part of the family, one of us, all together in this mad project that was my farming adventure. Surely the New Year would bring with it a tide that would change our fortunes and those lawyers would get off their backsides in the stylish manner that this pig had just demonstrated, and finally get the deeds signed.

Alice had now well and truly arrived, and with her she had brought responsibilities. We were her keepers and her life support. It was our duty to care for her, feed her and give her shelter. Perfect thoughts for a Christmas morning as we stood in that lowly stable.

If I could have read Alice's thoughts at that time, being in a somewhat sentimental frame of mind, as we were, I have often wondered what those thoughts would be. Possibly she had taken one glance at us as we leaned over that stable door, and thought to herself, 'Well, I've got that little lot exactly where I need them. Right in the palm of my trotters.' I wouldn't be surprised.

6 Sleeping Next to a Pig

As slowly as a thick fog lifts on a winter meadow, the lawyers finally did a deal and we moved into our little farm. It was early spring, and what better time to be enjoying the lengthening days, the sight of grass bravely beginning to grow, the return of the dawn chorus, and the subtle smell of rural rebirth to be enjoyed at every turn. But it was not long before we were enjoying a dawn chorus of a very different kind. There was no trill of the blackbird to wake us, nor any cooing from the usually immovable pigeons who took flight at the first hint of the unusual early morning sounds that now reverberated round that farm. The truth is that pigs don't go in much for sentimentality and pay little heed to the wonders of nature and the magic of the passing hours. They sense the sunrise and want feeding, there and then. They want it now!

Perhaps it was a mistake to choose as our bedroom the one that was nearest to the pigsty. It was less than a hundred yards away with nothing but open space between us. A separation of a couple of miles might have been better, or a well-placed brick wall. Tell most people that and they recoil at the possibility of the stench, but that wasn't the problem; it was the noise. If Alice had been living in an upstairs flat and you were unlucky enough to be the one living underneath, you'd be calling the police and taking out a restraining order. 'Turn it down Alice! For God's sake!'

It wasn't just the grunting or the snoring, which were amiable enough; it was the way she kept moving the furniture around. Despite the fact that there was very little by way of home decoration in this cosy sty with a floor space ten foot square – very generous by modern pig-keeping standards – she did an awful lot of interior redesigning. Nothing remained untouched.

Everything about a pigsty has to be built with strength. If you are contemplating keeping pigs, build their housing with all the robustness you can, and then double it. Sheep are easy to contain and the slightest bit of fencing will confine them as they simply don't have much strength, although they more than make up for it with stupidity. If you come across escaped sheep, it won't have been through effort but more likely idle opportunity. Pigs are far cleverer. They have necks with the latent power of a JCB digger, and being connected to their snouts means that snout can be turned into an impressive tool when it comes to doing a bit of DIY. So if they've decided something needs shifting, they'll work at it with the tips of their noses till it starts to give a little – it might be a bit of brick, or a gate fitting – and after they have made some slight progress and achieved a modest result, they will set to in a serious kind of way, giving it every bit of effort a super-charged snout can deliver until the desired result is achieved. Pigs will demolish an entire wall if you let them, brick by brick. They won't charge it, or try to push it over (which they could) but they'll patiently work at it with that snout till the wall simply gives up. They'll certainly rip up a concrete floor given the chance. I've seen them do it, and they're quick about it; there's no leaning on the shovel for a tea break in the world of piggy demolition.

Alice once spent an entire night on an architectural project of her own devising. With her snout as her only tool, she loosened brick after brick, truffling at a sliver of mortar till it gave way, then another, and another, till one entire brick fell loose. Then she moved on with the patience of a prisoner of war laboriously digging a tunnel to freedom. Eventually, with a rumble, the final brick dropped and the entire wall that separated sty from run cascaded to the ground. Had Joshua not been available, Alice could have seen off the walls of Jericho with no trouble. She now had her freedom. And what did she do with her liberty? Nothing. She just sat among the rubble with a smug look on her face, as if to say, 'I can do what I like, and there's nothing you can do to stop me.' I think I realised from that moment that what we had on our hands was something more than a mere pig.

But civil engineering wasn't really her thing, so she turned to music. It was her heavy feeding bowl that was her instrument of choice. Remember, this was intended to be a traditional farm equipped with much that would have found a better home in a museum. In the case of pig feeders, which on a modern farm are automatic and deliver defined and pre-programmed dollops of grub, we employed heavy, cast-iron rings that only a strong man could lift alone. Into these we would pour slop, from a bucket, stirred with a stick. The centre of the ring was domed, and so anything that was thrown into it would slither into the trough. These were ideal for our method of pig feeding, which was to give our pigs a kind of swill made from, well, pretty much anything within the law.

Despite the fact that two of us usually took a side each when one of these rings had to be moved, Alice was able to flick hers around like a frisbee using only her snout to give it momentum. Although she couldn't do any real damage, the noise when cast iron fell on concrete was deafening. The structure of these feeding rings was not unlike that of a church bell, and since each was slightly different to the next in size and weight, they produced a scale of musical notes. Given enough pigs playing with sufficient feed rings, it might be possible to get through the entire 'Messiah'. Certainly, 'Jingle Bells' would not be the slightest problem.

Our problem was the proximity of these feeding rings, less than a hundred yards from our bedroom window. Alice did not allow her hunger to accord with conventional breakfast and tea times. When she wanted to be fed, she demanded it there and then – no questions. And if she didn't get her way, well, she could have taught a few divas how to have a hissy fit. It made no difference that it was four o'clock in the morning; she chimed away with her feed ring, toying with it as if it was a bell that would summon room service.

I remember those early nights on the farm as being sleepless ones. Not only through the anxiety of facing the cliff face of the learning curve which I had set myself but mostly because, for much of the night, Alice insisted on making her music. And nothing would stop her. I can hear those bells of hell, even now.

7 Persuading Pigs

My antique methods of farming were gleaned from equally aged textbooks, from which it didn't take long to appreciate that stock are vital to any self-sustaining system of farming. What we now call 'old-fashioned' farming is, in fact, a modern conservationist's dream with its lack of petro-based fertilisers, veterinary medicines and its near-perfect recycling of fertility. Of course, it produced less food and needed more labour, but that wasn't its fault. It is we who have changed our needs when it comes to producing food, in particular the need to feed us cheaply. But someone has to pay the full cost if the customer doesn't, and that has been the soil, the landscape and the livestock. All these things lead a far more pressured life than in the farming times that were my inspiration. Much of that wisdom recorded in Victorian times forms the basis of modern organic farming. In that sense, what we might call the 'traditional' farmer was way ahead of his time.

By grazing the land, livestock bequeaths a hefty lump of natural fertiliser at regular intervals to enrich more soil and grow lusher feed for more stock to graze. They even spread it for us. The one thing that would make me seriously happy, as a farmer, would be to have a rich, steaming manure heap of my own, billowing like an express steam train, and then plough it back into the land to make it alive and fertile again. The cycle of cropping and fertilising is one of nature's less appreciated miracles and to be a small part of it is a great satisfaction.

Alice was going to help me make that dream come true. Of course, one pig could not be expected to carry this task alone, even on her ample shoulders. Soon a husband would have to be found, and more pigs would have to be bred to share the burden. Spring was coming and with it romance would have to be in the air. I moved her from her winter quarters in the sty to the shade of the orchard, where we had built a shelter to protect her from the direct sun. On hot days, she only came out for a stroll in the cool of the moonlight when she resembled an obese witch's cat on the prowl. One afternoon, out of the corner of my eye, I thought I saw a dark, heavy thundercloud approaching and a promise of much-needed rain. But it was only Alice, wandering again.

The whole business of getting Alice to do what I wanted her to do, and not what she herself had in mind, became all-consuming. It turned into a moral dilemma and caused me sleepless nights. The question was this: when Alice had to be moved and would not budge an inch, how far should the farmer go down the path of persuasion and trickery before taking the fast-track of coercion?

Morally speaking, it would surely be best to rely on persuasion: no farm animal was ever killed by kindness so why not employ a soft approach? But there would inevitably come a moment when Alice would simply have to do as she was told and kindness would have run its course. What then? If I opted for force, was I going to scar our relationship forever so that from that moment onwards the sight of me would have her over the hills and far away, galloping off in fear? I didn't want this to be a master and servant relationship; I wanted Alice and me to be a team.

I was not at all keen to employ force because, if I'm honest, I knew it wouldn't work. As someone once wisely remarked, 'Never wrestle with a pig. You get dirty, and besides, the pig likes it.' In one early and naive attempt to bring her under control, I used a system of ropes I had seen described in an old farming book:

Tie three or four half-hitches above the fetlocks. Hind legs in a similar manner. Fasten a longer rope to that joining the fore legs and run it back between the hind legs ...

Well, Alice, who had her dignity, did not take kindly to having anyone fumbling around with bits of string near her rear end. I did manage to achieve some kind of parcelling arrangement and remember shouting, 'Come on old girl, we've got you now.' But she gave one slight shrug of her shoulders, the badly tied rope fell away, and she was out of it as speedily as Houdini.

When it finally became time to move her out of that orchard and on to new pasture, I wondered how Alice was going to play it. Was she going to push me to the end of my tether and test the limits of my patience? And would she store all this in her mind for further use so she could play this protracted and frustrating game of 'catch-the-pig' once again?

This led me to an inevitable conclusion – Alice-moving was a game of diplomacy. I suggested a direction in which she might like to go and hoped she took the hint. There was no point in prodding with a stick, for she would freeze. The game needed as many people as I could muster, each of them armed with a board; if a pig cannot see a way ahead, it will not go. I had learned that already. The day came and once my small army had gathered, I outlined the strategy and gave military orders: 'Use the boards to deflect her progress, then, if she heads the wrong way, stop her with another board and let her see only the direction in which you would like her to head. Understood?'

'Yes, Captain Mainwaring,' they replied.

Alice retained, of course, the option of standing stock-still whatever you did with the boards, but let us pretend that thought hadn't crossed my mind.

It turned out that at Alice-moving time any visitor was in danger of being called up and pressed into active service. It was unfortunate for our friend, Mr Thomson, the art dealer, that he happened to call one afternoon. Italian leather shoes which had only known the gentle tread of a Bond Street pavement, now found themselves up to the golden buckles in sodden pig litter. But Alice-shifting brings out the best in people and, rather to our surprise, he entered into the spirit of the thing – when the moment came, he slithered and pounced like a professional swineherd. Fingers that only hours before had been stroking gilded frames, grabbed the hind legs of Alice with a grasp usually reserved for a cheque. When he next raised a finger to bid at Sotheby's, few would suspect where it had been.

You can poke and you can shove,
But a pig he won't be druv.

(Trad.)

8

A Nose
for Food

In humans the nose is a mere dual-purpose organ, useful for breathing and filtration only. But that apart, it doesn't really do much at all apart from needing wiping or blowing now and again.

On the other hand, pigs have turned their snouts into the most versatile of organs, certainly in the farm animal world. Take Alice's lovely and eternally damp nose. As we've seen, she had already perfected its use in the construction and demolition business, but we were fast learning how to thwart her architectural ambitions by reinforcing everything around her till it would stand an atomic blast – though I remember that even that didn't dissuade her once from trying to, once again, knock through a wall in the night.

But by now she had expanded her range of use for that formidable snout and developed a finely tuned and highly receptive radar. Remember, her vision was limited to a narrow field by those large, flappy black ears, but she turned that to advantage when she realised that, rather than being a hindrance, they gave her what was effectively an old-fashioned ear trumpet. Once she'd picked up the faintest sound, such as the kitchen door opening, she knew something was happening. But she needed to know more. Who was it, and were they carrying any food? This was where the nose came in. It started to twitch, as if electrically alive. She raised her head and sniffed repeatedly, deeply, testing the air, trying to extract from it any hint that there might be a bucket of swill on the way. You could see the concentration on her face. Her nose became her principal means of gathering intelligence and nothing got past undetected. If it had been radar, it would have been military grade.

But there were aspects of that nose that were less helpful, to me anyway. It was not long before the once green, flower-speckled sward of the orchard floor turned into the muddy tumult of a wartime battlefield. With that snout she dug deep trenches and threw up steep banks of earth, these turning to cloying mud after a light shower. Sometimes she would work away at one spot, making a single hole deeper and deeper till she could almost disappear into it. She's depressed, I thought. She's digging her own grave.

What was she looking for down there? What lay hidden deep within that soil which could be of any interest to a pig? Possibly the roots of all manner of plants, tasty and full of succulence. Or there might be worms, which would surely add something – I'm not sure what – to her diet. There might be all manner of fungus-like morsels, for all I knew. Don't pigs traditionally dig for culinary exclusive truffles?

However, it's nothing to do with food, apparently. It's more basic than that. The real truth about why pigs –and especially sows, and certainly Alice – are so determined in their truffling is down to a fundamental sexual urge. There is something found in truffles, according to German researchers, that is also found in a male pig's testes. This works its way into the saliva of an excited boar, is smelled by the sow's ever-perceptive nose, and the reaction is to prepare the sow for the pleasures to come. The scent is 'musk-like'. And if a sow can't get it because there's no boyfriend around, she'll go digging for it. That's what made the pig such a good truffle hunter.

But the researchers didn't stop there. They found related substances in human males, particularly in sweaty armpits. Then an horrific thought occurred. Was this why Alice seemed so affectionate at times? Did she greet me kindly only because the effort of mixing her feed had brought on a sweat that caused all the creases in my body to emit the stink of fungus? Or perhaps I smelled like a parsnip, or a stick of celery, because these substances are found in those vegetables too.

Apparently, truffle hunting has gone to the dogs these days, and canines are the hunters of choice as their pawing of the ground does less harm to truffles than the snout of a pig. Also, the pig was more inclined to (typically) eat its treasure, whereas the dogs would leave it untouched. This may be the reason the German police abandoned their use of a pig as a sniffer-outer of illegal drugs – perhaps the pigs were scoffing them and getting high. It was the Hanover police who hired a sow called Luise claiming she was better than any dog because she would sniff around all day, whilst a dog would get bored with it after half an hour. Sadly, the police gave up the experiment after being the butt of too many jokes, which was a shame as they'd proved that pigs were less intoxicated by drugs than dogs.

Looking at my orchard after only a few weeks, I could have told them that pigs brought no subtlety to their digging. But the worrying thing that crossed my mind was that Alice might only be performing this one-woman cultivation out of sexual frustration. Were those craters only dug because of the depths of romantic despair? I moved the search for a husband for her up a notch.

But the digging couldn't go on, so much was obvious as the mountains grew and valleys became ever deeper. The only question was whether or not to employ the only deterrent that would stop this, and that would be a ring through the end of her nose.

The ring operates on the basic principle of turning digging into a miserable business. Pressure of the snout against soil causes the ring to grip the nose just a little bit tighter, causing discomfort, at which point the pig backs off. Some say the pig gets used to it and learns to master it, and anyway is still able to root around in undergrowth and through straw. Others will see it as a violation.

Knowing her now to be a cunning kind of a pig, I knew exactly what Alice would do. From her exalted position in our orchard, where the Bong Tree grows, she would wait for a passing owl or pussycat, be offered a shilling for it, and she'd beat them up to a quid before they'd had time to notice. That's how she would have got rid of it.

So we left her to truffle her way between the apple trees and decided that, at worst, we could always rent it out for training mountaineers.

9 Keeping Pigs – a Moving Experience

There is a document called a pig movement licence, or at least there was during my Alice-owning days. It was issued by the local authority, which required you to trudge all the way to their offices and declare yourself and your intended pig movement, for them to then approve. It was needed when taking sows to the boar, and vice versa, as well as when sending pigs to be killed. It has its purposes in that in the event of a serious pig infection being declared, the movement of all pigs could be traced and the disease confined.

What a pity there was no such document that taught you how to move pigs in the first place. Yes, the moving of Alice was on the agenda yet again.

Alice, as you will have understood by now, was a single-minded creature and had her own views about the direction in which her life should go. That method I have described of directing her one way or the other with boards to restrict her vision was fine as far as it went, but it was no more than a device for persuading her to deviate slightly from the course she intended to take anyway. And sometimes it took a long time, meandering up and down the fields when she need to be moved from one part of the farm to another (no licence needed for that, by the way, otherwise I would have had to take up full-time residence at the council offices as Alice clearly appreciated an away-day with a different bit of land to turn into a quagmire).

So we got out that battered old trailer and decided that time invested in getting her used to that was as good as money in the bank. It would be like having a dog, almost, that you could take anywhere you wanted.

I had already grasped that pigs cannot be made to do anything by force which they are not inclined to do. One chap told me to grab them by the ears while someone else got hold of the tail, and shunt them around that way. It never worked on Alice. She wasn't having any human grasp anything at her rear end, and certainly no human grasp could ever match her wriggling capacity, especially when coupled with a piercing squeal. The whole operation quickly turned into a titanic struggle. In the end I resorted to the tried and trusted method, exploiting her one fundamental weakness: her desire for food. Hence, the bucket never failed. I filled it generously with swill, laid it in whichever direction I wished her to go, and sooner or later Alice would follow. It might take five minutes or fifty, but sooner or later greed would overtake all other instincts and the reluctant hog would budge.

Commercial pig farmers would, no doubt, have laughed their socks off at the thought that any pig should be allowed any choice over the way it lived its life, but we allowed her an enormous amount of freedom, which gave me a lot of satisfaction. This once proud and independent farm animal, the pig, has seen its world shrink from the open expanses of the field, forest and orchard – its natural habitats – to the insulting confinement of the indoor stall where movement is restricted and sunlight never shines. In the most commercial set-ups, anyway.

Every so often during Alice's lifetime, this issue made headlines. Pressure groups and protesters made much noise, and governments whispered modest concessions in return. Things improved, a little, but the clock was hardly turned back by a few minutes when, in fact, it needed to be reversed by a century. The poor old pig had become part of industrial farming, and no one was prepared to close the factory and return it to a cottage industry.

In the midst of all this debate, I found myself in something of a unique position. What none of the parties had been able to do was speak to the victims. Because Alice and I had developed an ability to converse, after a fashion, I was able to ask her, straight out, what she thought about it.

At the time, she was residing at what might be called her town address. Normally she patrolled the orchard, which she considered to be her rural retreat, but she came up to the metropolitan bustle of the farmyard when it suited her. I can't remember why now, but it was probably due to the orchard having had its soil turned over so deeply that there was a danger of her approaching the earth's core. There was nothing for her to complain about, though. Her sty was not narrow and did not confine her in any way; it was roomier than many a million-pound London flat.

In cold weather she spent most of the day playing Garbo behind the door, emerging into the world only for sanitary purposes (always in the north-east corner) and for meals (taken in the south-west corner). Pigs are fastidious. She always lay in such a position that the damp, slimy extremity of her black snout peeped round the door. It twitched and scanned the air like early-warning radar. That way, she could detect an approaching swill bucket two counties away.

To get her view on factory-style confinement of pigs, I shut the door on her and bolted it. There was a dreadful shocked silence while she came to terms with the indignity. I sensed her delicate snout pressing against her door, but it did not budge. She grunted at it, but it did not surrender. Then an eerie silence fell, broken moments later by the faint sound of what I swear was a whimper. She *never* whimpered; it must be real distress. It certainly wasn't the squeal of anger she emitted when moved against her will, nor the sharp reproving snort she fired off when hungry for her meal. It was a cry. No heart could remain unmoved. I opened the door and let her out. Interview over.

So, let all pigs be free was my broad conclusion. But this idealistic approach is not entirely without its problems. On the sandy, Suffolk coastlands near our farm, farmers did indeed find it better to keep pigs out of doors, ranging free. They ringed the fields with electric fencing and scattered huts made of shiny corrugated tin, bent into a crescent shape with a simple wooden back and front. They looked like small Nissen huts and we employed them ourselves. Pigs loved them but – and here's the problem – some neighbours loathed them.

The issue of these pig arcs, as they were called, 'littering the countryside' made one nearby village kick up a stink more powerful than any pig could produce. 'They are turning our countryside into a silver city of pig arcs,' they moaned. Some were even confused enough to advocate that confinement of pigs and intensive production was the only way forward and we should import all our pig meat from abroad, ignoring the fact that this would mean that pigs elsewhere would suffer, but out of sight is conveniently out of mind. I thought I might bounce this one off Alice, but such a feeble justification for mass mistreatment of animals hardly seemed worth lugging across the farmyard, even in a swill bucket.

If ever there was a 'pig-in-the-middle' situation this is one. Nice country-loving, green-wellied liberal thinkers are repelled by the captivity of an animal as intelligent as the pig, but equally revolted by the sight of them when they are let free. 'It's not the pigs,' they say defensively. 'It's the silver huts they live in. They're just *too* ghastly.'

Well, having bought several of these pig arcs over the years, I must report that there was never a lot of choice. The Cath Kidston pig shelter is still some years away, and somebody's going to have to lump it. I hope it is not the pig.

Alice – a
10 Catastrophic
Pig

Beauty is widely assumed to exist in the mind of the beholder, but it might surprise you to know that mathematicians can place it elsewhere, according to a book I have been reading.

It had never crossed my mind that the great curvatures of Alice's body – and great they certainly were – such as the broad arc of her torso that ran the length of her belly to the perfect curl of her tail, or the flowing mass of her hams running from the breadth of her back down to the tips of her rear trotters, were anything other than an outline that had happened by accident. There's an old saying that a good pig 'should have the shoulders of a parlour maid and the buttocks of a cook', by which I assume they mean attractive, at least to a male eye.

So far my thoughts on the shape of pigs had got no further than my frustration at not being able to restrain one when I wanted to. With sheep it's easy; get your hand under the chins and point their noses to the sky and they're yours, for a while at least. Dogs on leads are no problem, even a bull in a proper halter is manageable. But with pigs I was having to take my imagination a quantum leap in another direction.

My thoughts on the shape of pigs were becoming ever more influenced by a book written by my friend Allan McRobie, a Cambridge engineering professor, who begins his ambitious book *The Seduction of Curves* by boldly stating:

> Curves have an attractiveness and an appeal that straight lines and square corners cannot invoke. Curves are the lines of beauty that speak to us on some deeply instinctual level.

That's it! What beauty would there be in a pig if it were the shape of a packet of butter? It was her curves that made Alice such a beguiling creature, not just that she was a pig. You could look at her from any direction you wished – from sideways, front on, back on, underneath (although I have not tried it) – and the beauty speaks for itself. G.K. Chesterton said, 'In short [the pig] has that fuller, subtler and more universal kind of shapeliness which the unthinking mistake for mere absence of shape.' It's true. Looked at without my thought or imagination, Alice was nothing more than a four-legged lumbering blob. But seeing her through fresh eyes revealed Alice to be curved to operational perfection. And here I must be careful and give a nod towards modern obsessions for correctness lest I am accused of appreciating Alice only for her looks. Let me state here and now that I admired her as much for her mind as for anything else. OK?

I cannot do justice to my friend's mathematical arguments on which shape gives beauty (indeed I can't understand most of them), but I am grateful to him for introducing me to the concept of 'catastrophe', in the mathematical sense. He explains that the simplest form of catastrophe is the fold, or turning point, which occurs in abundance in the human form and, together with other catastrophes such as the cusp and parabolic umbilic catastrophe, gives it attractiveness – think folds of skin. Catastrophes also mark points of change, and he explains that these 'curved and spiky lines' form the boundaries of stability 'beyond which lie tragedy and disaster'.

I agree that pigs are indeed a walking catastrophe; that much I was learning. And it is true that on an almost daily basis the keeping of pigs is but a heartbeat away from tragedy and disaster, as you will have gathered by now.

But pigs are not just good, if catastrophic lookers. They can do sums as well. The 'Learned Pig' was a sensation in the late 1700s, and lived in London where it showed off its intellectual capabilities to excited crowds, having previously travelled the length and breadth of the country, to great acclaim. Apparently supporting the pig's intellectual credentials, Samuel Johnson wrote:

> the pigs are a race unjustly calumniated. We do not allow *time* for his education, we kill him at a year old.

The implication being that if pigs were allowed to live longer lives, who knows how clever they might become. In fact, the Learned Pig had been taught to pick up a series of numbered cards in response to instructions from its handler and could solve arithmetical problems apparently in its own head. With such mental agility now so evident in the pig, it caused some speculation as to where it all might end. The poet, Thomas Hood, wrote:

> In this world, pigs, as well as men,
> Must dance to fortune's fiddlings,
> But must I give the classics up,
> For barley-meal and middlings?

So if the pigs were allowed to live to a ripe old age, the argument goes, they might have had long enough to work out what my friend discovered, which is that their entire lives are shrouded in catastrophe. Which might explain why Alice spent so much time in quiet contemplation. She was looking at herself, and her shape, and figuring it all out.

11 A Palace for Alice

Did I mention that Alice was pregnant when she arrived? It's usual, when you buy a sow, for it to be 'in pig'. It's like the gift that goes on giving. Buy one pig, get another dozen free. The only difference this time was that she had not been the victim of a running service, where any kind of assault may have taken place at any time, but that there had been a proper introduction to the boar, a period of flirtation, followed by a consummation. I hadn't seen it but the boar's owner apparently had. In fact, I didn't ever want to see it. I'm not naturally squeamish but it doesn't sound like my idea of a good night out.

For a start, the boar will start to piddle up trees or walls, depending on what's available – so far, this could be a Saturday night in any UK town you care to mention. Then he produces a foaming saliva that bubbles round his mouth, and then he gnashes his teeth. So far, so alluring.

Don't worry, it soon gets romantic: sow and boar start to nuzzle each other and he'll give her a shove now and again, to see if his luck's in. If it is, he will mount her for a session that lasts for thirty minutes. Yes, pigs have sex for a full thirty minutes.

As I say, thankfully, I saw none of this. Except the boar keeper told me, 'he's stocked her well', so I imagined he looked on in rapt attention for the entire half hour.

From my point of view, Alice's pregnancy meant that she would need special housing, as befits a matriarch. I called it a Palace for Alice.

Clearly, the arc of corrugated iron that had given her shelter in the stable wasn't going to do the trick. It was just a bit too far from the house to keep a good eye on her, there was no electricity to help keep the piglets warm, and it was draughty if the wind was in the wrong direction. In all, it simply wasn't good enough. Having said that, they're cheap to buy and thrift always has a certain appeal to a farmer, even an amateur one.

The best housing ever designed for a pig must have been the cottager's sty. It consisted of two parts: a low, fronted house in which the pig rests and sleeps, and an outer run where she comes to feed and exercise. Dimensions are hard to define, but the run should be at least three yards square or she is going to feel cramped. The shelter need be no more than eight foot square. We were lucky to have a fine little row of sties, traditionally built, and to me they looked perfect. I hadn't given them much attention when Alice used one as a temporary home when she first arrived, but on closer examination I saw that more effort had been given to the detail of their construction than to any of the other farm buildings, even the stables where the much more valuable horses live. Clearly, pigs were held in high regard back then.

I suppose in an ideal world your pig would have two residences; a summer quarters and a winter retreat. A simple shelter for summer use in an orchard or patch of woodland, and when the autumn rains made the ground so boggy that all the soft soil lifted by your pig's snout turned to sticky mud, you could bring her back to the cosy and cleaner confines of the sty. That's how I figured it anyway.

But there's another important reason to keep your eye on a pig when she is farrowing, or giving birth. Pigs are generally good mothers, but big, fat, ungainly sows have a habit of rolling on piglets and killing them. It is not deliberate, but I was warned that a hefty pig in a hazy and exhausted post-farrowing state might fling its body mass around in a fairly uncontrolled way. It is a miracle that more piglets do not succumb to their mother's steamroller tendencies.

One of my bibles, wherein may be found more ancient wisdom that I could consume in a lifetime's farming, is *Stephens's Book of the Farm* of 1884. It is a magnificent work which was compiled with great attention to detail by this Scots farmer, and so authoritative was it that it became a standard work of reference throughout the Victorian and into the Edwardian era. There is much of it, especially its principles of natural husbandry, that are relevant to this day.

Of pig housing, Stephens wrote:

It is a great advantage to have stout battens fixed along the sides of that part of the sty on which the bedding is laid. The battens require to be 1 1/2 to 1 3/4 inch thick, and from 4 to 6 inches broad, depending somewhat on the strength and nature of the wood. They should be firmly fixed with their under surface from 8 to 9 inches above the level of the floor. Galvanised iron tubing 2 1/2 inches in diameter may be used instead of the battens, and is considered better from a sanitary point of view, but the iron is cold. This arrangement is a useful protection to the young pigs, as they can creep in between the mother and the wall and obtain a share of the maternal warmth without running the risk of being overlaid.

Let the building begin. As I understood it, the wooden rail provided a total exclusion zone; a solid barrier behind which the piglets can lie without any fear of mother coming and resting her great bulk upon them. It made sense to me and a stout length of timber was ordered, a few bricks removed to accommodate the ends of it, and then cemented in place. But it crossed my mind that the piglets might not recognise it as a place of safety, so I hung a warming infrared red lamp, hoping it might lure them like the loom of a welcoming pub on a wet night.

I spread deep straw to provide a decent bed, and we persuaded and cajoled Alice to descend from the heights of the orchard and into the sty. She was now in the maternity ward, the Lindo Wing of the pig world, fit for a princess to give birth.

And then started the waiting game. And we waited … and waited.

How do you know when a pig is going to deliver its piglets? Are there signs? I was acting on the advice of an elderly neighbour who told me to observe her nipples carefully. This was quite tricky when she was standing, but easier when she was asleep, which was often. Then, the nipples were on proud display and, feeling as if in the grip of an indecent perversion, I took a long, lingering look at them.

The trouble was I didn't know what I was looking for. My neighbour, on the other hand, did. 'She won't be having them little-uns yet,' he declared with great confidence. I asked why? He replied as if it was the most obvious thing. 'She ain't appled up yet!' Apple up? Didn't mean much to me. Where was this apple supposed to appear? He pointed to her udder. 'Yer know what I mean?' he asked. I shook my head, and then he cupped his hands. 'Appled up. She ain't appled up yet.' I got it. Tits as big as apples, he meant.

Applein' up is a guaranteed sign that the sow is producing milk and the piglets are on their way. I was getting seriously anxious by now because I felt certain the piglets were imminent, but the signs were still uncertain. The applein' up had only reached the stage where her udder looked like a bag of Cox's Orange Pippins when what I wanted to see was a sack of prize Bramleys. Then I would know for sure that some action is imminent.

A couple of nights later, by which time I had given up all hope and decided she was having me on with all this pregnancy business, I woke from a deep sleep with the sound of ringing in my ears. I assumed the stress of it all was beginning to show.

The chime rang out again. It was half past two. Wide awake now, I flew to the bedroom window. In the moonlight I could just make out the pregnant shape of Alice making frenzied music with her feeding bowls.

As I have mentioned, pig troughs are no lightweight affairs: they are cast-iron rings which take strong lads to lift. But Alice has been blessed with the power-packed snout that by now I knew only too well, and it was nothing for her to slide her muzzle under one of these hefty troughs and, with a flick of her head, heave it into the air. When it came down to earth, spinning, it sounded like the very bells of hell. From the bedroom window I loudly advised Alice to cut out the Quasimodo impersonation and tried to go back to sleep.

Of course, pregnancy does funny things to women, and pigs. Next morning I found that, as well as revising her dining arrangements, Alice had also done a thorough spring-clean of the sty, moving the clean straw out into the sun and leaving the grubby stuff in a heap near the spot where she dunged. 'Daft old pig,' I muttered into her floppy black ear, pouring her breakfast into the relocated trough.

A few hours later, we had eleven piglets. Just like that! Magic. First there was nothing, and then in no time at all there were eleven shiny black creatures, looking a bit like wet kittens, that slid from their mother with the greatest of ease, shrugged off their natural cling film wrapping, and staggered in the direction of a nipple with a determination that brought a tear to my eye. It was the bravest little journey in the world. It all took place nonchalantly, out in the sunshine on the clean straw. There was no fuss, except the commotion I made as I ran to tell the children. 'There's two!' I cried. Then ran back to the sty. Then back to the house. 'Three!' I sprinted from farmyard to house bringing news of every birth. By the end I was panting more than Alice.

I rang the owner of the boar to tell him the good news and he was delighted. I remembered what he'd said after I'd picked her up. 'I'd say he'd stocked her well, my old boy has.' He certainly had. She was as well stocked as a supermarket at opening time. In fact, when it got to ten piglets I started to get nervous and wondered if it would ever stop. 'Er, Alice, you can … er … stop now, thank you.' As soon as she was back home I marked the calendar. Pigs have a convenient gestation period of three months, three weeks and three days. We now know that the happy union took place on their first night together. It's lucky that black pigs can't blush.

There was a brief crisis when one piglet got caught beneath its mother's bulk as she turned. I was tempted to dive in and help, but as soon as the little one shrieked Alice rolled the other way. It was the only moment when I thought I might have to play midwife, which was just as well as I had been rather dreading it ever since I read a 1920s book that said:

> There are few problems in farrowing that cannot be solved by good humour and a plentiful supply of lard.

But we needed neither. Alice did it her way, unaided and with great dignity. She had done us proud.

Over the following years, Alice was to give us many litters of fine little piglets, though not all of Alice's farrowings proceeded as smoothly as her first.

One drama started at midnight on the longest day of the year – Alice always had a strong sense of the theatrical. The day had been hot, the wheat field was aglow with red poppies. It had been the best of summer days.

Just as I was getting into bed I heard a squeal from the pigsty, so faint that it could easily have been a slight movement of a rusty hinge. Except that it had a blend of bewilderment and frustration that I had, by then, learned to recognise as the alarm call of the newly born piglet. Alice was at it again.

She rarely made a mistake in giving birth, so I was slightly concerned at the whimpering: Alice's confinements usually took place in dignified silence. I dressed and went across to the yard and soon discovered the problem. Generally speaking, you can bet that no sooner are piglets born than they set off purposefully for the nearest teat, which they find with hardly any trouble. It is one of nature's miracles. But one flash of the torch into the dark sty revealed Alice's major miscalculation. No doubt in order to enjoy the cooling breeze around her rear end while giving birth, she had plonked herself down in the doorway with her back end in the fresh air but her nose in the warmth of the sty. It was quite a wide door but she was a very wide pig, and consequently when the newly born set off in search of something to suck, they found themselves impeded by a firmly wedged Alice. It was like expecting babies to cross the Black Mountains for their first taste of mother.

I employed a technique not used in obstetrics in this country for some time and shouted 'Get up, you daft bitch!' So shocked was this grande dame at being addressed in such a manner that she heaved herself onto her feet, ambled inside, and settled down again without even bothering to give me a grunt. That pig had a withering way with her silences.

Next morning we had eleven piglets and another major problem. Alice had a teat the size of a cricket ball, and just as hard. It felt hot and looked tender. The vet confirmed infection and warned that other teats might be suspect too. He gave her an antibiotic but doubted whether Alice would be able to feed her litter. Only nine teats among eleven piglets spelled trouble. Although it was a weekend, the vet drove off to find us a sack of 'Sow Milk Replacer'. But when I saw the gloomy package and its list of contents, I recoiled. It read like a food additives horror story. The 'milk' had antibiotics, growth promoters and a sinister ingredient described as being 'denatured according to EC regulations' (aren't we all?).

I am not against giving drugs to animals to cure or to save lives, but I hardly felt these healthy but hungry piglets deserved a pharmaceutical belt round the ear at this early stage in their lives. So I discussed it with my wife, and we came to the conclusion that Alice herself might conceivably hold the solution to her own problem. We went back to the sty and told the old sow that we were placing our full trust in her and we decided on a course of careful and dutiful observation. We paid hourly visits, sometimes bringing bunches of fresh clover from the meadow, like relatives visiting the sick. Clover to a pig is as good as a bunch of grapes to your granny.

Due to the influence of her medication, Alice was clearly under the weather. She didn't move much except to eat, and then only with half an appetite – a state I never, ever thought she was capable of. As for the piglets, we expected one morning to find two of them dead and some kind of natural selection to have taken place, but decided that 'Alice knows best' was our policy. Despite all predictions, a week later all eleven were thriving. I do not know how she did it; perhaps she had devised a rota system, for none of her piglets looks underfed or sickly.

Alice soon regained her strength and was soon back on full milk-bar duty. She was a heroine whose determination saw her litter through. Alice was a mother in a million.

12 Alice Has Her Cake and Eats It

I knew very little about how to feed pigs at first. I remembered there was something called pigswill, and that when I was at primary school we were severely warned not to drop forks into the slop bucket where the leftover school dinner went, in case a pig swallowed it. Apparently, all this discarded stuff went off to be boiled before being fed. It was a neat and productive method of recycling which, needless to say, has since been outlawed. If they invented the idea today it would be hailed as an environmental breakthrough.

Eventually, I was able to grow some of the pig feed on the farm itself, but Alice had arrived long before that. So I went to the local feed supplier and asked him what he recommended for a sow. He told me there I had a choice of two different feeds. I asked the difference. Without missing a beat, he replied, 'One is medicated. The other is *highly* medicated.' The magic medication was an antibiotic which, somehow, stimulated growth to get pigs fatter more quickly so they could be killed sooner. I certainly wasn't going to take a step onto that treadmill.

I was perfectly clear in my own mind that to feed an old-fashioned breed of pig, you needed an old-fashioned diet. I chose barley meal and middlings, sometimes known as sharps. Middlings are a by-product extracted during the milling of wheat, and they feel a bit closer to dust than does bran, and rougher than the flour. They make fine feed for pigs, and ours did well on middlings.

A farmer I knew, who kept a herd of commercial pigs, was persuaded by his family to fatten a Tamworth sow for their own consumption. He mistakenly fed it on the same high-growth, high-protein, medicated ration that he gave his commercial farm pigs. The sow surely did grow, but when it arrived at the butcher's and the carcass was split, the fat was sufficiently thick to see an entire family of Russian peasants through a Siberian winter. The meat was non-existent. The feed was too rich; the pig turned to fat at the expense of meat. So, a modest mixture of barley meal to sharps (middlings) in the ratio of two to one seems to work for a traditional pig. Some would say there is too little protein in this mixture and they may be right. Certainly, I found that litters got to a certain size and grew no larger even when I increased the ration, and this may well have been a protein deficiency. So, in addition, I gave a little mineral mixture in each feed, derived from calcified seaweed that was treated with sticky molasses to give it sweetness. This recipe looked so good once I'd cooked it up that I could have fancied it myself.

And then there was the greatest treat: the mangel-wurzel. It reads like a made-up name from a kid's comic, but mangels were once a basic farmyard feed-stuff given to pigs, sheep, cattle and horses. Farmers even used to make mangel wine out of it. If you haven't seen one, and they are rarely found these days, they have the look and colour of an overgrown radish. In a good year they will be as big as a rugby ball. The means by which they provide nutrition is a mystery, for they contain nothing more than sugar and water. As one old farmer once remarked, 'Mangels! They're 99 per cent water, but it's *bloody* good water.' And it is. I have yet to meet a farm animal that won't happily settle down with a mangel like a child with an ice cream cornet.

They are quite troublesome to grow, and doubtless that is why they are out of favour. From each seed can spring four or five individual plants, which need thinning to singles or the crop will be poor. The only way to do this is for a farmhand to work his way along the row, swishing back and forth with a hoe, employing the deadly accuracy that is needed to avoid killing the entire crop. Then, when they are ready for harvesting, they have to be picked one by one and carted to the farmyard and built into a stack, then covered in straw to keep out the frost that will destroy them. Bloody good water perhaps, but bloody hard work to produce.

It was very tempting to give the pigs any old kitchen waste, but I held back. Clearly, feeding them their own meat is asking for trouble, but even modern processed foods are too complex a cocktail to be innocently hurled at a pig. Left-over ice cream might contain unknown animal fats and this is how disease can spread. Cabbage leaves, eggshells, cooked potatoes, milk and cream always seemed safe, even though (strictly speaking) they weren't within the rules.

But rules are meant to be broken, and on children's birthdays there was always leftover cake. Alice, being the head pig, always got first go at it. It was solemnly carried from the kitchen to wherever she happened to be and placed before her. There is no more joyful sight than that of a pig, eyes aglow, opening its mouth to its full width to devour an entire chocolate cake, complete with Smarties topping, in one single, glorious mouthful.

Bulls make money, Bears make money,
But greedy pigs never do.

(Trad. stock market rhyme)

13 Huffing and Puffing

Alice shrugged off all of the seasons, whether hot, cold, wet or dry. I remember one blazing summer spell when the drought finally broke and the rains came. Blades of grass sprang from the brown crusts that had formed over the meadows, limp green leaves on the mangel-wurzels awoke with new vigour, the kale stood to attention, and even the depressed turnips started to show signs of life. But it was the piglets for whom I was most pleased.

Having been deprived for most of their short lives of one of their vital natural functions, they were at last able to root. Until then the ground had been too hard: soft black noses, not yet made hard by the passage of time, made little impression on the concrete-hard soil. I remember how, when they were only a few weeks' old and still in the sty with Alice, she taught them to root in the straw. They would follow her example of bending the head and shuffling the nose forward with determination. If any of the youngsters failed to join the lesson, Alice would waddle over and press their heads to the ground until they got the idea. Pigs have a national curriculum too.

The ground had been so hard for so long, though, that I thought the piglets might reach maturity without ever having put these hard lessons to any use. They may have a short life to look forward to, usually about six months, but I wanted to ensure it was a good life. But how good could it possibly be without a full rooting experience? Yet, when the rain eventually came, their snouts were ripping up the grass in the orchard, just like Alice had taught them, tearing up the juicy taproots of the docks and generally making a joyous mess and creating hell for the worms.

Little pigs very soon turn into big pigs, and then arises the question of where they are to live when they leave their mother. Unlike kids these days who stay home till they are in their mid thirties, piglets need to move on. Anyway, Alice made it perfectly clear that after a certain age she wanted no more to do with them and the responsibility passed to me.

Thinking about where they would be housed, a story came to mind, the one about the three pigs; one built a house of straw, another of sticks, and the third – the wisest – built his house of bricks. The marauding wolf, you will remember, huffed and puffed until the straw and sticks bent to his violent breath. The brick did not yield, however, and that wise little pig lived happily ever after. The others made fine wolf fodder.

Had I taken the trouble to tell the litter of growing piglets that story, it might have instilled in their minds the value of property, and what happens to little piggies if their houses are less than sound. Then they might have treated their new home with a little more respect.

I have to admit that, despite all the advice in that famous fable, I had built my pigs a house of straw. They were hefty, tight bales of wheat straw, and it seemed a cheap and cosy little home. Even better, I thought I had done myself a favour because I had built it on a field that had grown potatoes. We had taken the crop but left behind were those small, damaged or slug-nibbled spuds that were not worth picking up. The thought at the back of my mind was that the pigs might clear them up for me; it would save me the effort and feed them at the same time. This two-pronged approach had great appeal. Anyway, I did not like the idea of pigs living on concrete if there was an alternative. So on the potato field was where I built their straw house.

I hammered stakes deep into the ground, stretched two layers of chicken wire between and stuffed straw thickly in the sandwich. I topped it off with a few odd ends of corrugated roofing tin and hammered a wattle hurdle across the entrance. The thick straw would keep out the chill winter winds that were on their way, but at the same time allow the house to breathe so there was no constant dribble of condensation, which is a common feature of a modern ark. It looked pretty too, and traditional. *Dun-Rootin*, I named it.

But I had a concern. When the pigs were older and the supply of spuds ran out, would they then direct their ever more powerful snouts in the direction of their home and destroy it? Would they develop an extreme eating disorder and start consuming their own walls? They might well, because if I were to throw a bale of straw at them, within minutes it would have been broken open, scattered as broadly as if a pitchfork had been through it, and swiftly eaten. How were the piglets to understand that the stuff on which they had been sleeping, and which they had been idly chewing, is no longer an object of destruction but vital to their survival? I admitted it was confusing for them. Imagine if you were used to sleeping on a sweet-tasting, edible duvet, which suddenly turned into the bedroom walls. Would you not be perplexed? For some reason I had this belief that pigs do share the human notion of home being where the heart is, and I hoped the litter of piglets would arrive at this vital understanding and leave the walls alone.

But knowing that pigs will always be pigs, and just to make sure they understood, I decided that every time I strolled to the field with the feed bucket, I would give them a line or two of the old fable, with a little embellishment along the lines of 'straw-building methods are much more advanced these day than they used to be' and 'structural integrity is more important than materials in modern construction methods.' I did not want them to think that building in straw is in any way second class. And anyway, if they wrecked it, I was not building them another one.

I'm sure piglets pass through that growing-up phase which in humans we call adolescence. Once they get to a certain age, all they want to do is lie around, grunt and eat. That was why I took exception to some work coming out of an American university where they'd examined adolescence in human beings and came to the conclusion that it 'has been a key ingredient in humanity's evolutionary success, and exists in no other animal species, not even apes'.

Wrong! I do not know which animals those professors had been studying, but if he'd spent a day on our farm he would have found it stuffed with animals that had turned adolescent behaviour into an art form.

The way it works with growing pigs is like this: for the first few weeks they are timid and coy, and if you entered the sty with the feed bucket, they whimpered at your approaching shadow. It's almost heart-breaking. Sometimes, for safety, they would press themselves into a piggy little pyramid in the corner. They were, without doubt, childish. But they soon grow out of that and within a couple of months they will be over their fear of the world and spend most of the day asleep, except at feeding times, followed by occasional exercise pottering around the field rooting for the odd worm. Nothing too strenuous.

Then it gets serious. Suddenly they are bursting with life, developing strength they do not know what to do with and now, when you open the door to feed them, they start to come at you like a platoon of determined Royal Marines. The damage they do! Snouts into every crevice in the wall, nuzzling away at the mortar, pulling bricks away for pleasure. OK, I thought. If they want to be real destructive adolescents, let's get on eBay and see if I can find them a second-hand bus shelter to wreck.

Much as I love pigs, there were times when I would have happily seen the backs of them for good. But the moment I sensed that things weren't quite right with them, another instinct kicked in. I suppose it was all part of my becoming a stockman who needs to be able to be indifferent to one's animals one moment, and then bestow on them all the care and attention they might need the next.

I remember one cold night when the rain was turning to sleet, darkness was falling, and I decided I would feed the pigs before the weather got worse. The instant I got to the sty I sensed something was wrong. Those piglets knew the sound of the kitchen door, even though it was a hundred yards away, and even before I got across the yard they would usually be squealing as if suffering from morbid hunger. But this time they were silent, deadly silent. Not a sound. No hint of enthusiasm, no excitement, no squealing as they battled each other to be first at the trough.

My first thought was that they were feeling cold and keeping their heads well tucked under mother's warm and ample belly. But when I poured the swill and still the frenzy failed to materialise, I got worried. Alice lumbered to her feet and took a few mouthfuls, just to be polite; but the litter of black piglets remained heaped on each other in a bundle, breathing deeply, wobbling like blackberry jelly. Slowly they stirred, and my worst fears were confirmed: those curly little tails, which should normally be coiled as tight as bedsprings, were hanging with a depressing limpness. As the rugby song has it:

When a pig's a failure it straightens out its tail – but all pigs' tails are curly, 'cos piggies never fail!

These pigs were sick. Failing fast.

The vet declared the fast-growing litter had a feeding problem, so I gave them all an extra dose of minerals. A day later, the piglets' tails took a distinct turn for the better. Everything on a farm, even me, was better when coiled as tightly as a spring.

14 Alice Needs a Mate

I shudder to think how much time I spent worrying about my pigs' marital relations. It can get quite complicated when you've got more than one sow and several bloodlines running at once. It gets even more tricky when you're dealing with rare breeds, like the Large Black, because you have something of a duty to keep the bloodline alive. Inbreeding has doubtless been the downfall of many a breed of farm animal and we didn't want it happening to our beloved Alice and her family. There wasn't even computer dating back in those days, which might have made things easier. So the only way of finding Alice a boyfriend was to make yet more muttered phone calls, like a desperate father seeking to cast his daughter off the shelf.

It was worth the effort, though, because Alice always seemed to look forward to her journeys to her various husbands, and began to associate climbing the trailer ramp with making whoopee. Once she had made that connection we never had any trouble getting her into it; she scampered up the tailgate with the enthusiasm of a promiscuous teenager boarding the bus into town, intent on pulling.

But when a suitable marriage couldn't be arranged, things could turn very dangerous. If Alice came on heat and there was no boar in sight, anything male would do instead. I discovered this the hard way. It is considered efficient pig-keeping to send a sow back to the boar two days after she has farrowed. It all seemed indecently soon to me, so I often let her have a bit of a holiday before getting back to the serious business of churning out piglets once again. I hadn't realised the dangerous personal position this would put me in. Have you ever been propositioned by a pig?

As soon as she came on heat for the first time after giving birth, the litter of piglets was forgotten and romance filled her horizons. The sight of old Alice frisking, barking, skipping and sniffing the air to see if a boyfriend was within scenting distance was like seeing your maiden aunt leaning against a bus shelter and whistling at the passing sailors. Feeding her became a very tricky business. Having detected no boar in the vicinity, she assumed I must be the next best thing. You may take it from me that being wooed by several hundredweight of black pig is more unnerving than a wink from Mae West. Ignoring the food bucket, Alice rubbed her head violently against my trouser leg, opening her mouth wide and squealing coyly. I ran. She ran after me. She had the look of a woman intent on getting her man. I threw the bucket at her and had the fence not come between us, what followed might well have made anthropological history.

Having survived one of Alice's equatorial 'heats', I decided that I would not risk being a victim of the next. I urgently rang round the local breeders to see who might have a Large Black boar. There was not one to be found – at least, not one that was far enough removed from Alice's bloodline. So, I decided to conduct an experiment I had been planning for some time. I decided to allow Alice a mixed marriage with a Large White pig. Logic dictates that such a match would provide grey pigs, but in fact they turn out spotty like Dalmatians with areas of black hair amidst the white, all set on a pink skin. They also make, apparently, excellent bacon. If there is a snag with the Large Black pig it is that they are prone to produce copious amounts of fat. This was good if you were a peasant on a desperately poor diet, less good to someone who saw hideously lean bacon as the ultimate.

Fellow farmers were swift to recommend this union of black and white. 'Put her t'a large White' was the advice from an old pig farmer who came to gaze at the slumbering sows. 'I did bacon for Walls in the fifties. Grand bacon. Black pig with a white sow.' It was worth a try.

Now, it so happened that some friends had a Large White boar called Cyril. I rang and asked if he was free – not only was he free, he was willing. Cyril duly arrived by trailer just as Alice's next heat was rising from gas mark two towards the upper sevens. I asked my friends how they came by him, for he was truly a vast, pink beast beside whom Alice looked like a mere poodle. They told me he had been given to them by a pig farmer who wanted rid of him. 'He had trouble with his back leg. They were going to put him down but it seemed a shame.' I wasn't certain that a lame lover was what Alice needed; she was the sort of girl who demanded her mates be in full working order.

One other problem was that Cyril was a fussy eater. He lived in a backyard in Ipswich and had a convenient working relationship with the local bakery next door. Come closing time, any surplus jammy doughnuts and cream slices came Cyril's way. This made him the happiest, plumpest boar you could imagine, but his upbringing did not equip him for the kind of food that normal pigs ate. He had no appetite at all for barley meal and middlings, even with a mangel-wurzel on the side. He was of the fast food generation. When I poured out his slop, he gave me a distinct look that said, 'where's the custard slice, then?'

Alice and Cyril hit it off from the very beginning and soon they were sleeping side by side, looking for all the world like a sack of nutty slack and an overweight wrestler in fond embrace. Not many nights passed before I heard grunts followed by squeals of delight. It was embarrassingly loud. Love was undoubtedly in the air and I felt somewhat relieved that Alice's attentions were focussed on some other poor bloke.

Alice duly produced a litter three months, three weeks and three days later, as pigs always do. I sometimes thought she ran on clockwork. But for some reason, she wanted to give the impression that this latest litter was nothing to do with her. Every time a stranger approached, she ran to the back of the sty to distance herself from them, hoping people might think they were some other sow's litter, certainly not hers.

I think she was a bit ashamed. She didn't say anything at the time, and gamely shared the orchard with her flabby pink lover; but since those fruits of their passion came forth, she came as near to blushing as a black pig can. I guessed it was because her piglets were the most extraordinary creatures I had ever seen.

I had imagined we might end up with spotty piglets, like Dalmatian pups. But they turned out to have the markings of pink pigs dipped in ink, with no regular pattern to them. Some were pink all over except for a black bottom; others were smoky grey except for one bright pink ear. One looked like the map of Europe. But it was the shape of their ears which caused me most concern for they had taken after their prick-eared father and not their mother, whose ears, of course, are long and floppy. This put me at a great disadvantage, for the secret of my success in pig-husbandry was entirely due to the floppy ears covering the piglets' eyes; quite simply, they hadn't been able to see me coming. This made feeding easier, because they didn't mob me like feral kids at school dinner time, and catching them less of a race, for I could grab them by the back leg while their sense of smell was still working out where I was. But now they could see every move I made. I feared troubled times ahead.

Alice Needs a Mate

I am a pretty wench,
And I come a great way hence,
And sweethearts I can get none:
But every dirty sow
Can get sweethearts enough,
And I pretty wench can get none.

(Trad.)

True Love ~
15 Not Running Smoothly

I must admit I was quite pleased to see the back of old Cyril the boar, although Alice missed him greatly. It was not long, though, before she was once again her joyous, rotund, pregnant self. Every teatime, as hunger overtook her, she would stand on tiptoes on the edge of her trough and poke her black snout over the wall of the sty. Militarists would call it over-the-horizon radar. I called it a grub-seeking missile. Perhaps she was worried by the competition from the sty next door for here lived a new acquisition; a black boar of our very own, bought from another breeder to bring some new blood into my tiny herd. Technically a Large Black, he was in fact a Very Small Black, but he was only three months old, a real baby with not a hint of manhood about him. He was destined, though, to flourish into a strapping lad by the following spring, and then I planned to launch him on his career as the local gigolo.

Boars, even small ones, are to be taken seriously. Fully grown they can be terrifying, and tales of serious accidents are legion. It was therefore essential that, by the time the hormones started to pump through him, he was so easily handleable that I would not hesitate to take him for a walk through town on the end of a piece of string. That's how tame I like my pigs.

Much to my surprise, I had managed to tame most of our other pigs and the effort paid off handsomely. Alice would follow any bucket, even to the ends of the earth, and I would have no hesitation about taking her to a Palace garden party, providing swill and scones was on the menu. But I guessed our baby black boar was going a take a little more taming. For a start, he had come up from deepest Devon and hardly knew our strange East Anglian ways. The food was funny too: he had been used to processed pig nuts and now had to face the barley-based slop which we concocted. It is like being taken off cream crackers and put onto soup. I feared that re-educating him was going to be time-consuming.

We had a friend down from London who worked in an office where he said there were plenty of porky bores. The office had no windows, conditioned air and computer screens – intensively reared pigs will recognise this as the sort of place in which they spend much of their miserable lives. Our friend told me the latest management buzzword was 'face time'. You didn't go to meetings any more, you have face time with each other. So I decided that was what I would do. I would chat to my newly arrived pig, give him apples, and generally encourage him to feel part of the farm. Snout time.

The other problem I had was giving him a name. Anyone with any knowledge of pigs would appreciate the affection that goes into the naming of a much-loved hog. Pigs have characters that their name needs to reflect. I once named a young sow Thora after Dame Thora Hird herself, who came to our farm with a film crew to use our haystacks as a backdrop against which she could choose her favourite hymns. Did she mind having a pig named after her? She was over the moon. And so was the pig.

But I dare not even hint at some of the names that flashed though my mind for our young boar. Let it be sufficient to say that he was called Murphy simply because he was black and stout. My only hope was that he did not live up to his name entirely; we wanted no hint of bitterness in him when he grew up to be a big boy.

The course of true love did not run smoothly, though. Alice and Murphy, destined to be lovers, hated each other at first sight. What a disaster! She would have nothing to do with him. She snarled at him, struck him violently with her snout, and flung the poor boy away from the feeding trough as if she were tossing aside a mere trifle, and not a hefty lump of manly black pig. The poor thing could not have known what had hit him. I gave him credit for his perseverance, though. He kept coming back for more and he showed all the right instincts. Like a soppy lover, he followed Alice around with his nose so close to her backside that it looked as if there was a short string between her tail and his snout. The boy wasn't going to give up.

Still she took no notice. Things eventually descended to such a low ebb that one night I discovered he had been thrown out of the marital sty and was sleeping under the stars, while she slept in the warm comfort of the sty. There he lay for a little, grabbing a little sleep before regaining his spirit and fighting broke out once again. The row! It echoed round the farm, the anguished shriek from young Murphy as Alice started to beat him up again. It sounded like Jurassic Park out there.

Whether they finally got it together I have no idea, but three weeks after their supposed night of bliss, Alice came 'on song' again and it was no little madrigal performed in hushed voice. No, she came 'on song' like a Rhine maiden at Covent Garden. I had separated them, given the animosity they showed to each other, but I grabbed my chance and hastily shoved them back together again. I thought, hopefully, that time might heal. Does it hell! They were soon at it again; there were first coy squeals of joy – a good sign – but these were swiftly followed by shrieks of pain as they once again plunged their teeth into each other – not so good. Then silence followed, broken by a throaty roar. It was as if they had taken the entire emotional content of a long marriage and were acting it out every five minutes.

In the end I gave up. This match was clearly not made in heaven, or anywhere else on the planet for that matter. So I moved Alice away for a while to cool off. She went to a boar on the far side of the county and was away for Christmas. The farmer had promised me that he had a fine sire, but when I got there I was shocked to find him a lanky lad of hardly a year old. I am sure he had what it takes, technically speaking, but to see his juvenile frame alongside Alice's mature bulk … well, it was like expecting Pinocchio to woo Mae West. As I left them to get on with it, I thought, 'I don't suppose he'll know what hit him.'

16

Pig Sick

When you have kept pigs for a while, you arrogantly assume that you know their ways. That is precisely when they will take you by surprise. For example, it was only after several years of pig-keeping that I realised that pigs could, indeed, smile. Pigs' moods are usually easy to determine by listening to their grunts, which will be low and rumbling if in a conversational mood, or high-pitched and piercing if tetchy. But never in my experience did a pig have a facial expression. However, one Saturday night, I am certain I saw a pig grin.

We had just held the farm's annual potato-picking weekend, in which we invited the general public to follow our horses along the furrow to pluck the succulent organic spuds direct from the soil. Cunning wheeze, eh? Not only did it save me the trouble of having to pick them myself – which, with our aged machinery was a slow, back-breaking and expensive business (teams of pickers had to be paid) – but it gave an opportunity for those who have never seen carthorses at work, or plunged their hands into soil, to get stuck in. Children sank to their wellington tops in the mud and held contests to see who can find the largest potato. The tiniest children merely stood and gazed upwards at the carthorses, our mighty Suffolk Punches, and when they had come to terms with this overpowering presence, braved themselves to ask what the horses were called.

But the adults were the most entertaining. At any given moment you would see one man with a camcorder, who in his mind was remaking *Far from the Madding Crowd*, but in reality was not even likely to get it onto the nether regions of YouTube. Then there were the devotional green-minded ones who bent and picked with religious fervour, giving thanks for the rich smells of the freshly turned soil and handling each picked potato as if it were a miracle in itself. And we got others who muttered, 'My old dad grew taters. He wouldn't think much of these little 'uns,' as they flung aside a cannonball of a spud.

Back in the farmyard, a trusty team of neighbours manned the scales and herded the crowd towards our pocket-sized farm shop. Once there, they had to confront my wife who, it has to be said, is more at home at her writing desk than bending over a freezer being quizzed about the difference between rolled rib and topside. But she bravely played the part of the farmer's wife, only balking at questions such as 'Hey, didn't you used to be something or other before you worked here?' Strange how many people got a frozen hunk of brisket dropped on their toes.

So what was there in all this to cause the pigs to grin? The answer was the weather. Potato-picking, for those who are not being paid to do it, is a fair weather pastime. But the forecast was for torrential rain. I had decided that if it arrived as forecast, we wouldn't clear the entire field. And I certainly wasn't going to bother to pick the remaining potatoes – I'd had enough of spuds by now. Anyway, prices were so low that it was barely economic to unearth them in the first place. But they would not go to waste, for that most efficient of potato-lifters, the pig snout, would finish the job for me. I would turn the pigs onto the field and, if they wished, they could spend from now till Christmas in a nutritional treasure hunt. Somehow, word of this plan reached them and, as the storm clouds gathered, the piggies began to grin. I swear they did.

In the hope that the weather might not be as bad as forecast, we prepared for the second day, deciding to entice more pickers by offering cups of tea and a traditional Suffolk bun called 'fourses'. These are a heavy blend of flour, egg, lard and currants, described by my aged recipe book as 'filling'. It is said that ploughmen took them to the field on harsh winter days. The ones we made were better suited to propping up the legs of wobbly tables. Imagine scones cast in plaster of Paris and you get the picture. One old lady took one look and said, 'Mmm, I remember them. Of course, we were poor in them days. We had to eat things like that.' We had made two hundred.

To the credit of our customers, the weather did not deter them. In lashing rain they slopped along the furrows, wiping the mud from their hands on the grassy headland, applauding the horses who were having to pull with all their might to get the potato-lifting machine through the sodden soil. And as basket after basket was taken from the fields, the pigs grew ever more gloomy, seeing meal after meal disappear into paper sacks. They lined up at the fence and gazed despairingly as the last ridge was lifted. 'Never mind!' I cried to the pigs. 'It's fourses all round for tea!'

Not only can pigs grin. They can sneer, too.

They can also express deep displeasure. Alice in particular was good at this, and always let it be known when she was not a happy pig. Like a disgruntled monarch, all it took was one look from her. One day I got the message loud and clear, and it wasn't a happy one. 'Quite frankly,' sources close to her told me, 'she has had a bellyful of the press. She is pig-sick of them!'

Her privacy, apparently, had been intolerably intruded upon and she no longer felt able to settle her ample belly into a pleasantly cooling wallow of mud without some nosey parker peering over the hedge and asking, 'Oooh, is that Alice, the one that I've read about in the paper?' Alice was now featuring quite heavily in my weekly 'Farmer's Diary' in the Saturday *Times*. Things came to a head when one of these uninvited visitors produced a camera. In order to secure his exclusive shot, he cheekily rattled a feed bucket lying by the sty. Awoken from a deep sleep, she emerged looking far from her best: ears askew, eyes bleary, and tail hanging limply, only to find it was a false alarm. Shutters whirred. No doubt those pictures eventually fetched thousands from salacious French magazines. Where would it all end? How could Alice be sure that next time she went to the boar, the rat pack would not be hiding behind the feed bin to capture her private moments?

I knew she felt like this because, in recent weeks, Alice had been treating me with an increasing coolness. She had always been happy to use me as a way of channelling her views to a wider world, not being in a position to speak publicly herself. But one Christmas, when I made private details of her farrowing public, I found myself cut dead. She rejected all tickles behind the ear as I poured the swill into her bowl. Moreover, sources close to her revealed that she was increasingly worried at the behaviour of the younger generation.

Phoebe, one of her many daughters, was just back from the boar and due to have her first litter that autumn. Alice feared greatly the publicity that would ensue, and wondered how the young girl would bear up under the strain. She was deeply disappointed that all pleas from her to me were met with an (admittedly) arrogant, 'It is in the public interest.'

There the matter might have rested, had not this great family of black pigs been brought together by near-tragedy. Phoebe suffered a physical collapse. Her back legs alarmingly ceased to work. She could drag herself along on her front elbows in the manner of a seal, but that was as much as she could manage. Rather than ring the vet, I rang another breeder who told me he had had a similar experience and was advised by his vet that there were only two solutions: one was an injection that would cure her but abort her piglets, and the other was to shoot her. I didn't fancy either, and neither did my breeder friend, whose solution to a similar problem was to turn his pig onto the meadow and see what happened. I did ring the vet, in the end, and he didn't think that was too bad an idea either.

By day six I was worried. She had not risen. She was not in any way dispirited, or off her food. Indeed, she was extremely chirpy, for a pig. But still she did not rise. I feared the worst.

Then, in a second-hand bookshop, I found yet another aged farming tome written for novice pig keepers back in the 1920s. It advised that pigs that fail to rise should be given brewer's yeast and cod liver oil. It was worth a try. Off to Boots I rushed and within the hour I had administered the dose. Ten minutes later – I swear it – the prostrate sow had risen on all four fat black legs and walked. The Lazarus Pig. We all rejoiced. Alice forgave me all my transgressions and in return I vowed that, when I put pen to paper in the future, I would be more discreet.

But although Phoebe went on to have her litter, and several after that, she was not destined to live the full and long life that her mother, Alice, so clearly enjoyed. My suspicions were aroused when, one night, her feed bowl was not quite empty. Phoebe was probably the greediest pig we had ever raised, and her obsession with food was such that she would assume the slightest sound to be a harbinger of fodder. Certainly, it was not possible to open the back door of the house without her beginning to squeal, even if she had only been fed twenty minutes previously.

A pig that does not eat is truly a sick pig, and so it proved to be. Having known her as a robust animal all her life, it was sad to see the life ebb from her, and it was equally confusing for her litter of piglets who could not understand why the ever-providing mother had suddenly run dry. The vet came twice and did his best but it was clear that she was not going to live. I knew an old horseman who swore he could take one look at a sick horse and know its chances of survival. 'Yer can smell death on a horse,' he said, but I hardly believed him then. I did now.

We gave Phoebe a grim little burial, the knacker-man not being interested in pig meat. Robert the farmhand and I took our spades and dug a grave in what was thankfully easy-dig, sandy soil. Our gravedigging had its lighter moments. 'Shall we drop her in Viking-style?' asked Robert, but I did not want to think of old Phoebe standing to attention six feet under. Instead, we dug a little deeper than we planned, fearful that we would find her trotters still waving in the air. And with the somewhat cheering thought from Robert that not many farm animals are lucky enough to die naturally in their sleep, our grim little funeral procession began. The cold rain, carried on a northerly wind, poured on us as we hauled the dead pig onto a sledge and dragged her up the farm. Her litter of seven did not even give her a backward glance as we pulled their mother out of the sty past them. Neither did Alice give her daughter even a nod of recognition as we lumbered past the orchard. Phoebe was a good old pig, a fine mother to five litters, and we missed her.

Why a Pig
Will
17 Never get Lost

Pigs never cease to amaze me with their talents far beyond my wildest expectations. Take their ability to navigate. Under the heading of 'isn't it wonderful what they can do these days?' comes the personal satellite navigator, which were an astounding innovation at the time I was farming. For under £300 (yes, they used to be that expensive), this pocket-sized device would plot your position on the planet to within a hundred yards. It does not matter if you were halfway up Everest or in the queue for chips, it would tell you precisely where you were and – even cleverer – would give you detailed courses and directions to get you back where you started from. It is difficult for us children of the 1950s, who thought television was a marvel, to believe that a handheld machine can actually track satellites and by measuring signals that pour down on us from space, the plucky little box can follow every movement we make.

Now, clever as all this might be, let me devastate you by telling you that, according to my observations, pigs have been doing this for years. I do not know how they do it, or from whence they get their signals, or how they do the sums; but without doubt the navigation system that pigs employ far outperforms anything pouring down on us from outer space. The satellite boys can offer us an accuracy of about a hundred yards, pigs have it down to a matter of inches.

This amazing observation came about as a result of the belated onset of autumn. It had taken some time coming that year, but finally the oaks had got round to shedding their fruit and our orchard was now awash with windfall apples and acorns. Send for the pigs!

Alice, well trained by now to follow a bucket of food, waddled along like a faithful dog for the hundred or so yards from field to orchard. On arrival, she sniffed, truffled a little earth with her nose which she usually does on arrival at a new home – her equivalent of kicking off her slippers – and settled down for the night, chewing on the acorns, determined to put on another layer or two of fat for the winter.

Polly, who was a younger sow and yet another of Alice's daughters, was not so easy about it. The memory of an old electric fence still haunted her. Having spent the entire summer carefully avoiding the strand of electric fencing which had confined her, there was no way she was ever going to cross the line where she believed it still to be. We took it down for a good half hour beforehand so she could get the idea it had gone but, despite all foodie temptations put before her, she would arrive at precisely the spot where her navigation system said the wire had been and would go no further. Not one inch. In the end, we waited for her to burrow her snout in the bucket, gave her a quick push till she was well over the line, and then she waddled on quite happily. But the most remarkable display of navigational skill was yet to come.

Polly had not been in the orchard for almost a year. She had been nowhere near it, had no sniff of it. A lot of swill has flowed under the bridge since she was last there. So explain this: as we came up the slope and through the gate, Polly came to a stubborn halt. Try to coax her on and she dug in her heels and squealed. Tempt her with more food and she would not be persuaded. She froze. We thought about this, wondered if her nose was detecting some strange animal scent of an enemy lurking, say a fox. Then we remembered.

It was on this precise spot, almost twelve months before, that Robert and I had struggled long and hard to load Polly into a trailer to take her to her sty for the winter. The *very* same spot, to within the inch. The problem then had been that, although the electric fence had been removed and replaced with a trailer, to get in the trailer she had to cross the threatening – though invisible – line. Eleven months later, and heading in the other direction, she remembered. Fantastic.

Which leads me to the conclusion that clever though satellite technology may be, it has got nothing on pigs. If ever I am on an expedition demanding accurate navigation, I shall certainly take a pig. They don't even need batteries.

18 Who Could Take Her Place?

*E*ventually, Alice's day would be done – I knew that. Even so, it was uncertain when that might be. How long does a pig live? It's difficult to get an answer to that question because pigs normally head for the butcher about six months old. Sows, of course, live long after that, but as soon as their productivity starts to fall off, they too will be destined for the sausage machine.

Nevertheless, I had to plan for the future. And I found myself choosing from our herd of young sows the one that would be the heiress apparent to much-loved Alice. Not that Alice's position as matriarch was in any doubt, but I had to make plans. So I started to cast an eye over Alice's daughters, and fine upstanding gals they were too.

I could say that with confidence, for I had before me a list of good points that a young sow should exhibit. Like a judge in a disgraceful talent contest where looks were the only thing that mattered – think Miss World in the 1970s – I took my research on the perfect body for a pig to the orchard where the four maidens were roaming. I started, as the National Pig Breeders' Association list suggests, by examining the head.

It should be 'well proportioned, broad and clean between the ears' (not as in the real Miss World, where blankness between the ears was an essential). Legs should be 'well set, straight and fine-boned'. Shoulders should be 'fine and in line with ribs'. The list warns against 'excessive jowl or undershot lower jaw' – presumably to keep out the inbred county element – and then we come to general movement. This should be 'active' – and now I'm blushing – the loin should be 'broad and strong' and the belly 'full, with never less than twelve sound, well-placed teats'. While I was trying to take all this in, the hungry young sows danced around the trough like flighty models on a catwalk. In the end I did it the way they did it in all beauty contests and went for the one with biggest features – more than twelve of them.

But it is one thing to nominate a future leader and another to accomplish the *coup d'etat*. The merest sniff of demotion and Alice was going to take it badly, I knew. Without arousing too much suspicion, I had to separate the chosen one from the rest of the herd, which would be bound, I'm afraid, for the butcher. I marshalled my forces. As a second lieutenant, I needed a tactician who had the measure of the enemy, so I sent for my old friend Dilly, a farmhand of fifty years' experience and a great comfort when moving stock, since there is no disaster that has not befallen him in that half-century.

We set up the gates and hurdles in the corner of the orchard, built a pen and started the long process of temptation which we hoped would corral three of the four pigs. We poured some swill and hoped they would be lured into our trap. They were, until they realised what was happening, and – with the forcefulness of the

SAS storming the enemy – exploded from their temporary prison. Hurdles went flying, pigs went scurrying. We picked ourselves up and started again.

Unfortunately, our contest winner was not enjoying life in the spotlight. The fact that the finger of fate had come to rest on her meant nothing. Whenever her three doomed girlfriends ambled into the pen she insisted on joining them. We explained to her that, given the butcher was waiting with his engine running, there was literally no future in such behaviour, but she took no notice. We had to pick apples from the trees for her before she was persuaded that out was better than in. As for the other three, greed got the better of them and, with several buckets of swill and old Dilly doing an impersonation of a pig so realistic I nearly put him in the pen as well, all three were safely captured and sent on their way.

Of course Alice knew nothing of all this commotion. She seemed to be enjoying life now she was rid of her greedy, brawling offspring. She had better get back to the boar and get productive again.

When the butcher's lorry trundles down the lane, you are forced to face the hard realities of farming, and set aside many of the wistful thoughts in which we have indulged ourselves so far.

We must admit that the only reason we keep farm animals is to eventually kill them; it is dishonest to believe otherwise. Don't worry, I don't mean dear Alice.

As you will have grasped by now, I had become unnaturally fond of Alice the Large Black Pig. She had such an intense character, coupled with an obvious intelligence, and had proved herself to be a careful and trustworthy mother, and I developed the utmost respect for this deceptively ungainly and unsophisticated animal. She was safe and could live as long as she liked.

Friends gasped when confronting the harsh facts of pig-keeping life. How could you send them off! Visitors, meat eaters all, happily leaned over the door of the sty, cooed at the sight of happy piglets suckling at the teat, but any mention of the butcher and they distanced themselves. They preferred not to give it a thought.

The care and loving attention that most livestock keepers bring to the task of raising farm animals is difficult to square with the idea that all that physical and emotional energy is ultimately directed towards ending their lives. If it worries you deeply, it is unlikely that you would have taken the first steps into farming in the first place. But it is self-deluding to think that this paradox has never crossed your mind. It certainly crossed mine. How can it not? What kind of morality encompasses the rearing of animals only to kill them is the question to which you keep returning.

The best answer, of course, is for food. We developed into the strong and enduring creatures we are only through learning to cultivate and grow crops to feed us. The eating of meat, the absorption of fat and protein, once the power of fire to cook it and make it digestible had been discovered, allowed us to develop even further, giving us strength to fully exploit later discoveries such as spades and forks, which gave us mechanical advantages. But these days, in less needy, less carnivorous and increasingly vegan times, these arguments seem out of place – except to meat eaters, of course.

The mistake, perhaps, is to elevate farm animals into something akin to an oddly shaped human; to believe they think and feel like us, have emotions and ambitions, feel depression or elation. In writing this way about Alice, I may have been guilty of this false portrayal, but I hope you will find it forgivable.

Pigs have instinct; I'm sure they do because all animals have, and out of that instinct might come a sense of danger and with it, fear. But animals have no moral code that we can discern. Certainly, in the case of pigs, I always assumed that if, for any reason, I stumbled and fell in the pig pen, if I wasn't on my feet pretty soon they would devour me without thinking twice. There will be no thought along the lines of 'he's been good to us and fed us, so let's help him to his feet' – fat chance. A sow will eat her own babies under certain circumstances and chickens will peck on the carcasses of a dead 'mate'. For all the fondness that I showed Alice, I knew in my heart that it was ridiculous of me to feel that it might be reciprocated. She was a pig! And I am a human, and it is an inescapable fact that, in the pecking order, I am higher up than she is. Those who want us to be treated as equals, and who wish the same moral code should apply, make the same mistake I do when I deduce from her actions thoughts she could not possibly be having. I love to imagine what might be going through her head, as I have done for the majority of this book, but I know it's nonsense. But that hasn't stopped my imagination from enjoying the entertainment she has given me, and I hope you as well.

Those who seek to stop all killing of farm animals live under the mistaken impression that farmers would instead be happy to have them wander around their fields for no purpose at all. Sheep are always used as an example. Don't kill them, they argue. Rear them for their wool and not for food. This displays a woeful lack of understanding of the pitiful value of a sheep's fleece. The last time I sheared a flock of sheep, the shearer charged twice as much as the fleece was worth. So we did it for the animals' welfare, and when you've only got a small flock you can afford to do it. But on a national scale? What happens to the twenty-three million sheep that currently graze the grass of the United Kingdom? The answer is that they will die out, disappear, become the object of well-meaning hobbyists who might keep a handful of them as curiosities. And what will we have done to the sheep by no longer killing them? We will have driven them to the verge of extinction – yet another paradox.

The knock-on effect of that, of course, would be a drastic change in the landscape. What we see as 'the countryside' is, in fact, a factory that farmers have created over countless centuries. The stone walls, the fells, the forests and glades, the farmhouses, the ditches and hedges – these are not crafted by nature but by man for the containment and feeding of livestock. These landscape infrastructures would soon crumble. The fells and dales would give way to bracken and brambles and the footpaths become overgrown. The same sadness that is felt on an obsolete factory floor would spread across the countryside.

If we stop eating farm animals, they would die out far sooner than you might think, and the countryside as we know it goes with them.

There is one argument put forward, and I am very taken with it, and it is this: a 'deal' has been made with farm animals and human beings. 'We look after you,' say the humans, and the animals respond by saying, 'and in return we will feed and clothe you.' I like that, as long as both sides stick to the bargain.

But I'm afraid not all modern farming lives up to the promise. Highly commercialised farming has led to some welfare atrocities that we need not outline here, but I once paid a visit to an intensive pig farm in Denmark, champion producers of cheap bacon. But what a price the poor pigs had paid. I was shown sows that had been bred to deliver as many as sixteen or more piglets at a time, rather more than Alice's modest eleven – a bumper crop and a good result for a farming accountant. Less good for the sow. In fact, by breeding these sows as 'super-producers' they had accidentally bred out of them the ability to deliver naturally. Instead of that perfect piece of nature that I had seen at Alice's first farrowing, these sows were delivering their piglets by open surgery conducted by a vet. Disgusting.

For the sake of our own consciences we must fight practices like this at every turn, assuming we ever find out about them. We have to respect farm animals during every second of their lives: protect them from predators; feed them to keep them healthy; and, when it comes to the end and we extract our part of the deal, we must make sure that this end is free from stress and suffering. If you can achieve all that, then there is no reason you cannot eat meat with a clear conscience. And that was what I strove to do every day of my pig-keeping career.

If you don't believe me, ask Alice.

Enough! Though you have ploughed your way through these philosophical arguments, when I sent my first pigs off to be killed it was a distressing experience, but not for the reasons you might imagine. It all started as a brush with the nineteenth century and ended up as a head-on crash with the confused values of modern times.

I always asked for neighbourly help when pigs had to be marshalled; capturing agile swine calls for a man of dogged determination, instinctive stockmanship and the ability on the part of the farmer to curb his tongue in front of the children. I had none of these qualities. Richard, my neighbour, had them by the ton. By chance, we also had a sculptor staying with us who claimed to have wide experience of pig handling, having spent time in the peasant cultures of mid France. I didn't really believe him.

Although I said nothing to them, for some while I had been wondering how I would feel when the first stock headed for slaughter. After all, these were pigs from Alice's precious litter. I had been with her on that sunny June morning when she effortlessly delivered them into the world. We had cared for them like babies, even thought of starting a photograph album of their piglet-hood – not quite, but you know what I mean.

To my surprise, I felt no remorse at their going. I can put my hand on my heart and declare that no pigs have had more comfortable, cosseted or better-fed lives than these. As the only purpose of raising pigs is for them to be eaten, I faced the abattoir with a clear conscience, with one provision: they must die as they had lived, with dignity.

For the moment, however, they were still free. They edged towards the bowl of barley meal I planned to use as bait, but sensed the hurdles designed to capture them were some kind of threat and the slightest twitch by any one of us made them flee. What a remarkable instinct they have for detecting a downward turn in their fortunes.

Even a six-month-old pig is unstoppable if it has made up its mind to be free, so when the sculptor advised and the neighbour acted, I let them get on with it. By macabre coincidence, a travelling pork butcher arrived with a wicker basket over his arm to inspire us with his hams, chops and sausages.

While the pigs' attention was diverted by the comings and goings of his van, we seized our chance and snapped shut the hurdles. We were halfway there. Recalling his Gallic adventures, the sculptor suggested we put their heads in the bucket of meal and, by applying gentle pressure, back them into the trailer. It turned out that he *did* know what he was talking about and minutes later we were bound for the butcher, only five miles away.

The slaughterhouse lay hidden behind a white-washed facade in a typical Suffolk village, although the slaughterman himself was not typical at all. He was the son of the former vicar, having inherited his job from his father. Vicar/slaughterman is an unusual conjunction of callings. He spared time to advise and sympathise with first-timers like me, like a godly man would, but more importantly the pigs got kind attention too. There was no stressful overcrowding or undersized pens: animals were killed within a couple of hours of being delivered. It was as far removed from an insensitive factory atmosphere as you could wish to be.

Needless to say, a few years after I started taking livestock there, this model slaughterhouse that worked the highest of welfare standards and gave farm animals the end they deserved, was closed by legislation which demanded that abattoirs which kill one hundred animals a week should be wrapped in the same bureaucracy as those that kill ten thousand. Surely, any fool could have seen that rules made to govern a steelworks will never work if applied to a blacksmith's forge, and neither would the rules of mass meat production ever have allowed small men, like my butcher, to survive.

The argument is long, but I know of no more sympathetic or stress-free end to a couple of pigs' lives than the one enacted in that tiny slaughterhouse. Is there the remotest possibility that the small and caring may ever be valued as highly as the mighty and efficient? Pigs might fly.

There was soon a domino effect, with one closure after another. The butcher in our nearest village finally put up the shutters and announced that he'd bought a cottage in France and was hanging up his cleaver for good as one more country butcher hit his own sawdust. His was a haphazardly built wooden shop, unsettling to any health inspector reared in the age of white tiles and stainless steel, but one that had given good service for two generations.

Although its meat was fine, it was the banter I missed the most. Whether you bought a bone for the dog or a brisket for the pot, Peter gave every customer at least a minute's quick repartee that set you up for the rest of the morning (and, come to think of it, induced you to buy something you never intended to). He told me he was trying to learn French. 'Trouble is,' the cunning old devil confided, 'I can't speak a word of it, but they'll only catch me once, won't they?'

I had been browsing for a few minutes when he tapped me on the shoulder. 'I have something out the back, you'll never guess.' It was a hand tool, with a wooden shaft like a small axe, except that on the opposite side to the blade protruded a blunt spike about six inches long. He proudly declared it to be his grandfather's old poleaxe. The poleaxe was a deadly device used for the killing of animals before the invention of the humane killer with its captive bolt. 'My grandfather, he could get hold of them bullocks and one swing with this,' he waved it threateningly in the air. 'And then they were gone.'

We went into the back of the sheds that had once been a slaughterhouse. Peter picked up a rusty and equally deadly looking cleaver. 'I can remember shining that till it gleamed,' he said, looking at the now corroded and pitted blade. 'Know what it's for?' He flung it in the air. 'It was for halving pig carcasses. You hung 'em up and one swipe of this and they were split in two.' I gave him a tenner for them both and, pleased with my purchases, put the cleaver and the poleaxe into the boot of the car, breaking my regular rule of never buying bits and pieces that I had no intention of using.

It was the small slaughterhouses, like the one from which I rescued the poleaxe and cleaver, that in a strange way were guarantees of animal welfare. All the villagers knew what went on; they sensed whether the place was clean or filthy and were their own environmental health inspectors. Equally importantly, they could judge the health of every animal that died. Cattle, pigs and sheep were probably driven down the main street, or at least brought in an open cart: there would be no chance of any sickly or mistreated animal escaping the gaze of the villagers. It would not have been possible for a cow exhibiting the pathetic symptoms of mad cow disease or something similar to pass unnoticed into the food chain. But, as we have learned to our cost, it can easily happen these days, despite shower after shower of rules and regulations. The rearing and killing of animals now happens behind closed doors.

Although I have been happy to send pigs to slaughter, safe in the knowledge that I have given them a decent and satisfying life, I doubt I would have the courage to pick up the knife and stick the pig myself.

19 Let's Give Polly a Ring

It is widely assumed that the owl and pussycat went to sea in a beautiful pea-green boat. But this is not true. If you really want to know what the owl and the pussycat get up to, you had to come and camp in my farmyard, any day at dusk, and keep as still as you possibly can.

The first thing you would hear would be a rustling amongst the newly spread straw in that part of the farmyard where the horses lived. This straw, I remember, was the remnants of a disastrous stack of oats that was devoured by hungry mice and rats before I had a chance to thresh it; but although the stack and its feed value were long gone, several families of mice seemed to cling on, like hopeful regulars sitting at un-laid tables in a favourite restaurant that has long since closed down.

Shortly after the rustling had commenced, you might have heard the creak of the stable door which led into the yard. That would be the black cat about his shady business. With great skill, he teased the door gently open with his paw and then, hairs bristling down the length of his back, crouched down for the first of several kills of the evening. The slaughter ended just before dark, when the distant sound of the tin opener rasping at the cat food tin lured him back to the house.

But the entertainment was far from over. Shortly after that, swooping low over the hedges and the ditches, would come the hunting barn owl. He would pause on a gatepost or a telephone wire and, when he had exhausted his usual hunting grounds, he'd circle the farmyard to ensure that all was quiet; if the barn doors facing the yard were open, he'd glide inside, perch on the beams, swoop for a tasty mouse amongst the sacks of corn, and feast for half an hour. Then, with a powerful beat of his gleaming wings, he returned to the nest. If you were very lucky, you would catch sight of owl and pussycat working together, as much a team as trained guerrillas, flushing out rodents and making the farmyard a better place to live. I cannot believe they would ever go to sea, even in a pea-green boat. They are simply too busy.

It crossed my mind that the owl and the pussycat might spare a few moments to have a word with Polly, our young, Large Black sow, and daughter of Alice. Again, it is widely assumed that pigs are open to offers for the rings in the ends of their noses. When invited to 'sell for one shilling the ring in the end of its nose', the piggy famously replied, 'I will.' So, do we take it that pigs hate their rings so much that even the slightest offer is worth taking up?

I mention this because I was faced with a dilemma. Should I ring Polly the pig, or not? Of course, as already explained, I would not have dreamt of doing it to Alice. She had been around too long, and was set in her ways, and should not have to handle such a major change to her way of life at her advanced age. But Polly was different. She had many years ahead of her – but if she continued to root in the way she was doing, I feared I may lose my control. It was a pity the contractors who dug Crossrail did not know about this animal. It only took a few stabs at the earth with that snout of hers for a substantial crater to develop. She was every bit as good at it as her mother. Often I thought she had strayed, only to find she had dug a hole deep enough to accommodate her entire ample frame and conceal it from view in order to continue her journey to the centre of the earth.

All this made a terrible mess of the land, brought subsoil to the surface, destroyed grass and anything else that was trying to grow. So I was contemplating ringing her, which would stop her from rooting and be in her own best interest, and prevent me from turning her into sausages. But would she understand? In the summer we had glorious fields filled with all manner of clover, lucerne, grass and other tasty fodder. I was quite happy for her to graze here, but grazing and digging are two different things. I simply could not afford to have my meadows ruined by potholes. But neither did I wish to restrict her natural instincts; there must be some good reason for the rooting and I would not wish to spoil her fun. Was education the answer? Could she be persuaded that there was pleasure in sniffing at pollen, chewing at dandelions, or snorting at butterflies?

To ring, or not to ring, that was the question? I had confronted this question early in my pig-keeping career but still had no answer. The arguments still echoed. Is a ring all that bad? I would have hated her to live in perpetual discomfort and I certainly did not want to crouch in the farmyard one evening as the sun was going down, only to find the black cat making her a cash offer for it that she would be all too willing to accept.

After much soul searching I came to an inevitable conclusion; in her Christmas stocking this year, our young sow Polly will find a shiny, golden ring. But that was months away and we were now at the stage where she could shift more concrete in a night than a motorway worker on double time. It only took one tiny crack to appear in the floor of her sty and she would start to sniff it gently, snout twitching in anticipation of a glorious night's excavation. Having decided there was scope for further earthworks, she would worry it a little more, and before many minutes had passed she would be raising slabs heavy enough to wind a champion weightlifter. She must have been rooting for something, but what? I have always assumed it was fat, juicy worms that pigs dug for, or crunchy roots that lie hidden beneath the surface. Or perhaps they were bored, in which case it would be reasonable for a little mischief to break out. But even when she has the distracting company of the boar – who is more than able of getting up to a little diverting mischief himself – Polly found time for a little vandalism amidst the romance.

As ever, I was grateful to all who offered me both advice and experience. One said there was absolutely no need to ring a pig; the rattle of a feed bucket will always divert them from their destruction. This adviser spoke after a wide-ranging, pig-owning career that included a Berkshire sow being brought home on the back seat of an estate car. Apparently, the sow sat up and peered out of the back window for much of the M25, but nobody noticed. Sadly, the pig developed a talent for 're-landscaping' neighbouring gardens whilst ignoring all rattles of feed buckets, and that was that.

It was the actual insertion of the ring that caused me some concern. I knew that in the case of ringing a bull it was usual to call the vet, but with pigs it is the farmer who was expected to do it – I had no idea where to start, and I suspected that these sturdy rings didn't come with instructions. I remember a journey on the London Underground where I noticed that the wearing of rings in the nose (by human beings) seemed to be fashionable. At least, I assumed the girl was wearing it for fashion reasons; I cannot imagine that she was inclined to root up concrete for the fun of it. Sadly, she left the train before I could get close enough to see if the ring was stamped with a country of manufacture, or ask her whether she had inserted the ring herself or got the vet to help.

I had almost made up my mind to go ahead with the ringing when a fellow farmer told me of Percy, a boar now of blessed memory, who dug 'sitting holes' for himself. As he became ever larger, so the craters became deeper, and for the sake of the field it was decided he should be ringed. Percy, by now a manly chap with a mind of his own, seemed none too keen and it took two terrified vets and dose of tranquilliser to accomplish the deed. But the ring did not stay in place for long and, after Percy had got through four of them and no vet could be found with the courage to try a fifth, the farmer gave in and let Percy have his way.

I was deeply suspicious that my attempts to ring Polly would go the same way. She had that 'go on, I dare you' look in her eye. Anyway, Christmas was coming near, so I told her that on the fifth day, my true love would send to me five gold rings. If she couldn't leave the concrete alone, she would get the lot of them.

What Is It
20 About Pigs?

We can only assume that, when they held the contest to choose man's best friend, the pig did not bother to enter. Otherwise she would have won. But that would be typical of a pig, to prefer the sanctity of a muddy wallow, or the idleness to be enjoyed in dappled shade, to the sordid demands of public performance. Pigs are private animals and it is only when you have kept them on a small scale as I have, that you are allowed glimpses of the way the pig organises its highly civilised world for its own comfort. We must always remember the wise saying, attributed to Winston Churchill, that a dog looks up at you, a cat looks down, but a pig looks you straight in the eye. I take this to mean that there is something about pigs and ourselves that is enacted on the same stage. This, perhaps, is the reason we find them such objects of fascination.

There is certainly no point in thinking you own a pig, for they are quick to remind you who is in charge. If they do not like where you have put the feeding bowl, they toss it to a preferred position with one deft flick of the snout; if they are not enjoying the confinement of the pigsty, they work at it with a deadly combination of nose and teeth till it is as they wish it to be. It's very similar to what we try to do as we do our best to tunnel out of lifestyles we are not enjoying.

Pigs are certainly quick to put you in your place and remind you that, in the matter of giving birth, they have far better ideas than you do on how things should be managed. One of our sows, I remember, made the apparently fatal error of giving birth to thirteen piglets while having only twelve teats to feed them. As each piglet is assigned its own teat for the suckling period, this appeared to be extremely bad news for number thirteen. The vet's advice was that the runt should be bottle-fed, or put out of its misery.

But we allowed the sow to run things her way, and observed over the following hours how she nudged the runt into life and taught it to dive for the first teat that became free, so that it could mop up the leftovers its bigger, bloated siblings could not manage. We used to watch it making its way along her belly like a desperate drunk at closing time, supping the dregs.

But, as I have said, pigs are not pets; they are farm animals kept for food. They have bravely stood between families and starvation for centuries. Because they could be fattened on kitchen and farmyard waste, killed and cured and made to last through a long winter when no other meat could be preserved, pigs were the foundation of cottage economy. Now that they have become part of a semi-industrial process, not only are their lives demeaned but the food they provide us with has deteriorated to a poor imitation of what pig meat in all its glorious forms used to be a mere fifty years ago. The sooner the words pig and industry are divorced the better, and then the pig can lead the life it deserves. If we are prepared to pay the price of producing meat that way, then we will eat the better for it.

But this is no satisfactory answer to the question, what is it about pigs? We know they have been faithful providers for centuries. But so have cows, yet I doubt they are held in such close affection as pigs. To be honest, a pig has little obvious beauty. They are impossible to cuddle, devoid of fur, seem to enjoy filth and have the table manners of, well, a pig. But there remains *something* about them.

Ask anyone from the late Archbishop Runcie, a supporter of the Berkshire breed; to the Prince of Wales, who gave money to ensure the survival of the Gloucester Old Spot. Add to that distinguished list the fictional Lord Emsworth, and the generation of kids who fell in love with 'Babe' from Dick King-Smith's book. And there was the furore over the Tamworth Two, a couple of pigs that became national heroes when they made a daring escape from the slaughterhouse.

So why have we fallen in love with them on such a grand scale, when we only have to look backwards to find a time when little affection was wasted on them? Two hundred years ago, the first edition of the *Encyclopaedia Britannica* described the pig as a 'fat, sleepy, stupid, dirty animal, wallowing constantly in the mire'. The much later *Children's Britannica*, on the other hand, says of the pig, 'Pigs are intelligent animals ... they are also naturally clean.' Some makeover in the intervening years there.

The pig started so low in public appreciation that the high esteem in which it is presently held is all the more remarkable. Pigs, as Robert Malcolmson and Stephanos Mastoris (authors of *The English Pig: A History*) point out:

> … served in part to define in consciousness a boundary between civilised and uncivilised, the refined and the unrefined.

In a letter to his son, written in 1749, Lord Chesterfield wrote:

> Remember that when I speak of pleasures, I always mean the elegant pleasures of a rational being, and not the brutal ones of a swine.

This was an era when all things piggy were terms of abuse, and expressions like 'pig-headed', 'hog-wash', and 'pig's breakfast' were words of deepest condemnation. A Victorian book, written for children, could hardly cast the pig in a role less like the one she currently enjoys; the children who read *The Champion Pig of England: A Story for Schoolboys*, were warned:

> The career of our hero, Grunter Growler, resembles in several respects the existence of many two-legged animals who wear out their life in idleness and,

avoiding work, dedicate themselves to meals, sleep, and basking in the sun, so that this history may possibly teach some of us useful lessons.

Pigs had few distinguished friends throughout a long, dark period, except for Charles Darwin's grandfather, Erasmus. It is possible that a wider understanding of animal species ran in the family even then, for Erasmus wrote in the late eighteenth century:

> These animals, which are esteemed so unclean, have also learned never to befoul their dens, where they have liberty, with their own excrement; an art, which cows and horses ... have never acquired.

But Erasmus Darwin had foresights even Dick King-Smith has only just caught up with; two centuries before King-Smith penned *The Sheep-Pig*, later to become the hit film *Babe*, Eramus wrote:

> I have observed great sagacity in swine, but the short lives we allow them ... prevents their improvement which might possibly be otherwise greater than dogs.

But, while most intellectuals heaped scorn on the pig at every opportunity, its true friends recognised its enduring virtues. Alas, the pig's greatest-ever friend was also their ultimate foe. The country labourer knew only too well that the 'cottage pig' was the main-stay of his economy, and no matter how well he fed and cared for her, or leaned over the sty and scratched her back with a stick, he knew that, as autumn turned to winter, the pig's days were numbered. Then the pig would return the compliment for a lifetime's care, and feed the family through the winter. It was observed in 1806:

> … the pig is in reality one of the best friends of the poor. He seldom fails to supply something for the table of the poorest cottager, at least, on Sunday. Every peasant has a pig.

And for the first time during this mutual dependence, sentimental affection starts to appear twixt man and his pig. Flora Thompson wrote of Victorian Oxfordshire:

> … the pig was an important member of the family, and its health and condition were regularly reported in letters to children away from home, together with news of their brothers and sisters. Men callers on Sunday afternoons came, not to see the family, but the pig.

To keep a cottage pig was to have a true appreciation of the animal, more honest than sentimentality. There were undoubted bonuses:

A dear old cottager of my acquaintance once showed me the accounts for a fat pig he had recently sent to market. They gave a credit balance of three shillings. 'Not much profit there, Stephen,' [the writer] ventured. 'No, there isn't, is there?' he agreed. 'But there! I had his company for six months!'

But even the efforts of the cottagers to appreciate their pigs can hardly account for the pig-fever that overtakes us today at the merest glimpse of a grunting creature. One writer, Stephanos Mastoris, had his own ideas, and he's not far wrong:

> They've got a close physical resemblance to teddy bears. Big bodies, small limbs. And in recent years they've been less and less seen, and in a way because of that their cultural presence has increased.

To put it another way, absence of real pigs has made our hearts grow fonder – as I know only too well, it is far easier to coo over them at a distance than to have to face the marauders every morning with the feed bucket.

More worrying, Mastoris quotes Chapter Six of *Alice's Adventures in Wonderland*, in which Alice saves a pig from 'murder'. He writes:

> The scene concludes with Alice ruminating on how some children 'might do very well as pigs'; in her

mind – and Carroll's imagination – the line between pigs and people was not difficult to cross.

Uncomfortable though this thought might be to those of us who have kept and killed them, this may be exactly what it is about pigs.

21 Thank You, Alice

When I think back to my farming days, it is always Alice who first comes to mind. It could just as easily have been the magnificent Suffolk Punches, the great carthorses of the eastern counties; or the shimmering ruddy bulk of our Red Poll cows. But it is always Alice's portly frame that fills my mind's eye.

I think it is not overstating it to call her a celebrity. Although she moved gradually in retirement, Garbo-like, frail and increasingly deaf, Alice the Large Black sow knew what it was like to see her name in lights. A few years ago, an extract from one of my *Times* columns was used in an advertisement for the newspaper. It was splattered along the walls of the London Underground, on roadside posters as high as a house. Suddenly, everyone knew of Alice the Large Black pig. People stopped me in the street and asked me how she was. Her health has been inquired of in Piccadilly Circus, and a message of love was sent to her by the guard of an Intercity train. It is the sort of attention of which the modern day 'influencers' of this world would be jealous.

Alice's life changed completely by being part of the farm, as did the character of farm for her being there. Whereas farm visitors would once have glanced at the pigs in the field and said merely, 'How nice,' now they wanted to know, 'Which one is Alice?' as they stood by the fence and repeatedly called her name. This brassed her off no end, for she was used to being called only for meals, and when she found that she was having to haul her considerable carcass several hundred wasted yards every day to satisfy a nerd with a camcorder, she let her displeasure be known by deciding never ever to come when called. As we were unwilling to indulge in a battle of wills, we frequently staggered across the field with buckets of her swill in case she starved herself to death, out of pique, like a starlet denied room service. But at least my mind was always at rest about one thing: anorexia was never going to be a problem. No amount of fame would put Alice off her food.

Alice died a natural death. We never sent her to the butcher, we never could. But having decided that, there arose the question of how long she might live in retirement. In the end, Alice saw seven glorious years, until the day I found her, stiff, in the orchard.

It felt as intense as a family bereavement. At such moments, it was not proper to talk of her failings, but if she is missed it will be in the autumn. This was when Alice was at her most cunning, and her most beguiling. It was when the apple trees were laden with fruit that she could be seen rubbing her great bulk against the trunk of the tree, till it produced a gentle shower of apples that fell at her feet. She would play this game for hours, enjoying the fruity rain as it fell from the trees, and then return to her second favourite occupation of truffling for worms. When such tasks weighed heavy on her you might think that the last thing on her mind was her family, but a piglet had only to squeal and she would be across to it with the speed of a black thunderbolt, checking it out before returning to the serious business of filling her vast belly. She died peacefully in her sleep and was buried beneath the apple tree she loved the most.

Before we allow too much sentimentality to slosh over us, there is a serious point to all this. Alice was allowed the sort of life which a pig would choose, if anyone could be bothered to ask it. There was no confinement in darkened sheds, no routine medication, no performance-enhancing drugs, no desperate pursuit of efficiency. So grieve not for dearest Alice, for she had the best of it. Instead, be sorry for all the sows who dreamed of a life like hers, and never lived to see it.

Grandfa' Grig
Had a pig,
In a field of clover,
Piggy died,
Granfa' cried,
And all the fun was over.

(Trad.)

About the Author

PAUL HEINEY is a well-known writer and broad-caster who discovered a huge desire to be a farmer – but not any old farmer. Having fallen deeply in love with the mighty carthorse, the Suffolk Punch, he decided that he would farm traditionally, with horses instead of tractors, and surround himself with the farm animals of that period. Amongst them was the Alice the Large Black pig, the inspiration for this book.

As well as a farmer, he is an ocean sailor, having taken his own boat to Cape Horn and back, and taken part in the gruelling OSTAR single-handed transatlantic race.

He has most recently been seen on television as a presenter of ITV's *Countrywise*.